SpringerBriefs in Complexity

Michael Golosovsky

Citation Analysis and Dynamics of Citation Networks

 Springer

Michael Golosovsky
Racah Institute of Physics
Hebrew University of Jerusalem
Jerusalem, Israel

ISSN 2191-5326   ISSN 2191-5334   (electronic)
SpringerBriefs in Complexity
ISBN 978-3-030-28168-7   ISBN 978-3-030-28169-4   (eBook)
https://doi.org/10.1007/978-3-030-28169-4

This Springer imprint is published by the registered company Springer Nature Switzerland AG.
The registered company address is: Gewerbestrasse 11, 6330 Cham, Switzerland

Preface

This book belongs to the science of science. The idea of this book appeared in 2007–2010 when I attended the interdisciplinary seminar of Prof. Sorin Solomon in the Racah Institute of Physics, Hebrew University of Jerusalem. The purpose of the seminar was to construct physical models of social phenomena. My long experience with Web of Science suggested me to look for citations to scientific papers and to try to model their dynamics as physicists do. The modeling of citation dynamics has been popular among physicists, and almost all such models were built by theoreticians. These models were quite general and mathematically rigorous but lacked proper calibration, namely, comparison to measurements was insufficient. I set my goal to build a fully calibrated and validated model of citation dynamics. To achieve this goal, I hoped to use my background and experience in experimental solid-state physics which should help me to design special measurements for model validation.

This book presents a stochastic model of citation dynamics which is based on the well-known copying or redirection mechanism and which was built using methods of network science. The combination of modeling and measurement revealed that citation dynamics of scientific papers is nonlinear. This nonlinearity has far-reaching consequences including nonstationary citation distributions, diverging citation trajectories of similar papers, and runaways or "immortal papers" with an infinite citation life-span.

This book presents a fully calibrated and validated model of citation dynamics. It can serve as a practical tool for quantitative analysis and forecasting of citations and impact factors. This book appeals to students and researchers in network science, citation analysis, and bibliometrics.

I am indebted to Sorin Solomon who introduced me into the wonderful world of complexity, supported me through all stages of this research, and induced me to write this book. I am grateful to Sidney Redner and Peter Richmond for their encouragement and advices.

Jerusalem, Israel
December 2018

Michael Golosovsky

Abstract

We consider network of citations of scientific papers and use a combination of theoretical and experimental tools to uncover microscopic details of its growth. Namely, we develop a stochastic model of citation dynamics based on copying/redirection/triadic closure mechanism. In a complementary and coherent way, the model accounts both for statistics of references of scientific papers and for their citation dynamics. Originating in empirical measurements, the model is cast in such a way that it can be verified quantitatively in every aspect. Such verification is performed by measuring citation dynamics of Physics papers. The measurements revealed nonlinear citation dynamics, the nonlinearity being intricately related to network topology. The nonlinearity has far-reaching consequences including nonstationary citation distributions, diverging citation trajectory of similar papers, runaways or "immortal papers" with infinite citation lifetime etc. Nonlinearity in complex network growth is our most important finding. In a more specific context, our results can be a basis for quantitative probabilistic prediction of citation dynamics of individual papers and of the journal impact factor.

Contents

List of Symbols

$A(t)$	Aging function for references
$\tilde{A}(t)$	Aging function for citations
a	Prefactor in the expression for the probability of indirect citations
$\tilde{a}$	Parameter of the $s(K)$ dependence
a_{mn}	Adjacency matrix
$B(t)$	Generalized aging function
b	Coefficient in the expression for the probability of indirect citations
$\tilde{b}$	Parameter of the $s(K)$ dependence
C	Total number of citations in a citation network
$c_{t,t-1}$	Pearson autocorrelation coefficient for additional citations
f_l	Fraction of the second-generation citing papers connected to their progenitor by l two-hop paths
$f_{uncited}(t)$	Fraction of uncited papers
G	The slope of the $\Gamma(K)$ dependence
$K(t)$	Cumulative number of citations of a paper after t years
K^{∞}	Longtime limit of citations of an individual paper
K_r	Onset of the runaway behavior
K_0	Initial attractivity in the preferential attachment mechanism
$k(t)$	Annual citation rate of an individual paper at year t
$k^{nn}(t)$	Mean annual number of the second-generation citations per one first-generation citing paper (the nearest-neighbor connectivity)
$M(t)$	Mean number of cumulative citations after t years
$m(t)$	Mean annual citation rate at year t
$m_{dir}(t)$	Mean annual direct citation rate
$m_{indir}(t)$	Mean annual indirect citation rate
$N(t)$	Annual number of publications in one discipline
N^{nn}	Mean number of the second-generation citing papers per one first-generation citing paper
$n^{nn}(t)$	Mean annual number of the second-generation citing papers per one first-generation citing paper
$P(t)$	Probability of indirect citation through a paper t years old

P_0	Probability amplitude of indirect citation
Q	Community size
q	Reduced probability of indirect citation
R_{0i}	Reference list length of paper i
$R_0(t_0)$	Average reference list length of the papers published in year t_0
$R(t)$	Age distribution of references
$R_{dir}(t)$	Age distribution of direct references
$R_{indir}(t)$	Age distribution of indirect references
$r(t)$	Reduced age distribution of references
$r_{dir}(t)$	Reduced age distribution of direct references
$r_{indir}(t)$	Reduced age distribution of indirect references
s	The number of two-hop paths connecting a second-generation citing paper to its progenitor
s_0	Intercept of the $P_0(s)$ dependence
$T(t)$	Memory function for indirect citations
t_0	Publication year of a paper
α	Exponent characterizing the growth of the number of publications
β	Exponent characterizing the growth of the reference list length
$\gamma, \tilde{\gamma}$	Exponent of the memory function
Γ	Obsolescence rate
Γ_0	Obsolescence rate of the low-cited papers
δ	Exponent of the aging function for direct citations
ϵ, ζ	Exponent of the preferential attachment
η	Paper's fitness
Θ_{ikj}	Probability of indirect citation: paper i cites paper j through the intermediate paper k
λ	Probabilistic citation rate
λ_{dir}	Probabilistic direct citation rate
λ_{indir}	Probabilistic indirect citation rate
μ	Mean of the log-normal distribution
ν	Exponent of the power-law distribution
Π_{ij}	Probability of citation of paper j by paper i
Π_{ij}^{dir}	Probability of direct citation of paper j by paper i
Π_{ij}^{indir}	Probability of indirect citation of paper j by paper i
π_l	Probability of indirect citation in l-multiplet
$\rho(\eta)$	Fitness distribution
σ	Standard deviation of the log-normal distribution
τ_0	Citation lifetime

Chapter 1
Introduction

Abstract We explain what is citation analysis and which role it plays in the research of power-law statistical distributions, complex networks, and bibliometrics.

Keywords Bibliometrics · Power law distribution · Complex networks · Citation analysis

1.1 The Place of Citation Analysis in Science

Science is an evolving network of researchers, projects, and publications. Citations are links that glue the whole network together. Studies of citation statistics and dynamics proved to be very important for several research fields. For example, bibliometrics uses citations to build estimators of scientific activity and to identify research fronts. Power-law statistical distributions are very prominent in bibliometrics and historically this was one of the first fields where these weird distributions were discovered. The emerging field of complex networks also originated in the studies of citations of scientific papers (Fig. 1.1).

1.1.1 Bibliometrics

First scientific publications were handwritten letters and books with no formalized style. The emerging scientific societies, such as Royal Society and Academy of Scavants, established scientific journals with formalized style of correspondence, in such a way that the letter of private correspondence had been replaced by a scientific article. Initially, the scientists gave credit to previous works in the body of the letter or article. With the growth of science and the concomitant increase in the number of references to prior studies, these references became placed either in the end of the text or in the footnote. The role of references was twofold: the credit to prior work, on the one hand, and the means of retrieving scientific information backward in time, on another hand.

© The Author(s), under exclusive license to Springer Nature Switzerland AG 2019
M. Golosovsky, *Citation Analysis and Dynamics of Citation Networks*,
SpringerBriefs in Complexity, https://doi.org/10.1007/978-3-030-28169-4_1

Fig. 1.1 Citations are at the
core of several research fields

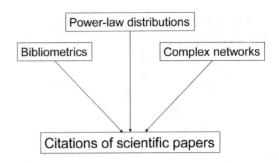

In 1950–1960, Eugene Garfield realized the necessity to trace development of science forward in time [58]. To address this issue, he organized the Science Citation Index. Initially, this database was used mostly for indexing and tracing scientific information but soon it found important application for the article-based or journal-based scientific estimators, in such a way that the number of citations acquired by a paper became an indicator of its popularity and quality, while the journal impact factor became the measure of journal importance.

Citation count got a boost in the information age with the appearance of electronic databases such as Web of Science, Scopus, Google Scholar, etc. For individual researchers, the role of citations as a tool to retrieve information forward in time has decreased in importance because electronic search using keywords can do it better. The role of citations shifted more to science mapping and to citation-based estimators of scientific activity, of which the Hirsch index [79] was only a harbinger. Recently developed alternative measures, which are covered under the term altmetrics, such as readership statistics (Mendeley), and Twitter, added more tools to this pool of estimators [173].

These developments enabled the citation analysis to become a part of the research field known as scientometrics. The latest advances in this field have been summarized in [6, 112, 116, 147, 163]. To name a few applications of citation analysis:

- Mapping of science.
- Identification of research fronts and hot topics.
- Scientific estimators, impact factor, quantitative research assessment, etc.
- Paper's quality assessment.

First two applications heavily rely on the structure of citation network and community detection. These applications are beyond the scope of this book which focuses more on citation dynamics and its relation to the structure of citation network. Our results are most important in relation to scientific estimators and prediction of citation dynamics of papers and journals.

While there are many empirical estimators of paper's quality that work well for the papers published in the same year and in the same journal, it is not clear how to compare papers published in different years and/or in different fields. To this end, one needs quantitative, model-based estimators. Existing models do not take adequate account for the fact that citations form a complex network. To understand microscopic origins of these estimators and to find how they evolve in time, we need a solid base for comparison between different research fields. To this end, a good model of citation process is required.

Another issue, for which we need a better microscopic model of citation dynamics, is the paper's quality definition. Indeed, citations became the currency of scientific estimation. Although everybody agrees that the number of garnered citations measures something important about the paper, it is not clear how to elaborate a uniform measure of the paper's quality. In their assessment of the quality of the paper, the editors consider the list of attributes such as novelty, timeliness of the results, clarity, significance, originality, etc. Since these attributes are not qualitative, they are subjective. The time came to elaborate the model-based definition of the paper's quality. This can be better done using models of citation dynamics and this will allow comparison of papers of different age and from different disciplines.

1.1.2 Discrete Power-Law Statistical Distributions

Another field where citations of scientific papers traditionally played a very significant role is the fat-tailed statistical distributions [41]. While Gaussian distributions dominated scientific thought of the nineteenth century, the advent of modern statistics in the beginning of twentieth century brought highly skewed statistical distributions to the forefront of scientific knowledge; the first examples of these were income [129] and city size distributions [4]. Then there appeared three bibliometric laws accounting for scientific productivity [103]; bibliometric use of journals [19]; and word frequencies in texts [43, 188]. These three laws were the first to introduce discrete power-law distributions and they were followed with the power-law distribution of citations to scientific papers, established by de Solla Price [45] and Redner [142]; and power-law degree distribution in the WWW and other complex networks established in the seminal works of Barabasi-Albert [8] and Huberman-Adamic [81]. All these distributions are fat-tailed, their tail exhibits a power-law dependence, $p(x) \sim 1/x^{\nu}$ where $\nu \leq 3$. Such distributions lack either a well-defined mean or variance.

The origin of these discrete power-law distributions still remains obscure. Citations to scientific papers is one of the first and best-documented examples of these weird distributions. By establishing the origin of the fat-tailed citation distributions, we can better understand their occurrences in other contexts.

1.1.3   Complex Networks

The field of complex networks traces its origin to de Solla Price who was the first to realize that scientific papers form a network in which citations are links that glue the network together [45, 138]. He also understood that much can be learned not only from citation distributions but from the structure of citation network as well. This idea was used later by Henry Small to identify research fronts and scientific communities through cocitation analysis and bibliographic coupling [156].

With the proliferation of electronic databases and the appearance of the WWW, it became clear that complex networks with highly-skewed degree distribution are not restricted to citations but are ubiquitous in nature [7]. Most of these networks are dynamic and the important question arouse: what is the mechanism of their growth? Although de Solla Price already in 1976 suggested the cumulative advantage mechanism to account for the growth of citation network, this study remained unknown beyond information science. It was the seminal Barabasi-Albert work [8] that introduced preferential attachment—a universal mechanism of the complex network growth. This mechanism, which is very similar to Price's cumulative advantage, became the paradigm for conceptualization of the dynamics of complex networks. It also became the basis for development of the realistic models of network growth.

To validate the network growth models, the measurements on real networks should be performed. Citation network, which remains the oldest and best documented complex network, became a playground for experimental verification of various models of network growth [53, 155, 185]. Such models were developed by computer scientists, mathematicians, and physicists. While the models of citation dynamics developed by computer scientists are mathematically simple and, in most cases, reduce to linear regression, they contain too many empirical parameters. Although these models are quantitative and calibrated, they are empirical and do not give much insight into the process of citation accumulation. On another hand, while the mathematical and physical models of citation dynamics are more sophisticated and contain smaller number of empirical parameters, they are conceptual rather than quantitative, and most of them await for proper calibration and validation.

1.2   The Purpose of the Book: A Quantitative Microscopic Model of Citation Dynamics

After bibliometrics has accumulated a lot of empirical knowledge on citation dynamics, there appeared a lot of models aiming at organization of this knowledge into coherent framework. While existing models give a clever insight, they are partial, non-calibrated, and capture only certain aspects of citation dynamics. A calibrated comprehensive model that captures all aspects of citation dynamics in one framework has been badly missing.

Traditionally, the physicists played a very prominent role in modeling citation dynamics (Shockley [153], Zener [184], Barabasi [8], Redner [142], Hirsch [79]). We describe here a physics-inspired fully calibrated and validated model of citation dynamics which contains a minimal number of empirical parameters. The novelty of our approach is in connecting citation dynamics to the structure of citation network. This is in contrast to existing models of citation dynamics [175] which, with few exceptions, do not account for network structure. Our stochastic model of citation dynamics is based on a well-known copying or recursive search mechanism. Our dedicated measurements aimed at determination of the parameters of the model and validation of model assumptions quite unexpectedly revealed that citation dynamics of scientific papers is not linear, as it was assumed previously. The nonlinearity has far-reaching consequences such as non-stationary citation distributions, diverging citation trajectories of similar papers, and runaways or "immortal" papers with infinite citation lifetime. We strongly believe that our nonlinear stochastic model of citation dynamics will be a basis for the quantitative probabilistic prediction of citation dynamics of individual papers and of the journal impact factor.

We start from a reasonable cartoon picture of citation process based in perceived author's psychology. Our model contains several empirical functions and parameters which are found from the calibration procedure. At the following step we verify the model, namely, we design measurements to validate the underlying assumptions of the initial model. Then we revise our model and perform second round of calibration and validation until everything is knit together. These are the subjects of Chaps. 2–5. In Chaps. 6–8 we analyze the consequences of the model and demonstrate the measurements that validate them. Other applications of the model, such as construction or justification of the estimators of individual scientists and community mapping, are beyond the scope of this book. In Chap. 9 we perform an analysis of existing models of citation dynamics based on our understanding of its microscopic mechanism.

Chapter 2
Complex Network of Scientific Papers

Abstract We consider scientific publications as a growing complex network where papers are nodes and citations are links which connect these papers together. We explain this network approach and make special accent on the temporal aspect of citation network, namely, we focus on the growth of the number of papers, age distribution of references, and citation dynamics. We trace relation of the age distribution of references to citation dynamics and explore this reference-citation duality.

Keywords Complex network · Citation analysis · Reference list · Growing science

2.1 Statistics of References and Citations

To study statistics and temporal evolution of citations of scientific papers, we adopt the complex network approach [7, 28, 121], namely, we consider the ensemble of scientific papers as a complex network where papers are nodes and the edges between them can be considered either as citations or references, since one paper's citation is another paper's reference. This is illustrated in Figs. 2.1 and 2.2. Citation network is temporal since each node and its outgoing edges (paper and its references) are created simultaneously; each node carries its own timestamp. This network is directed since each edge has a certain direction: from the citing paper to the cited paper. This network is acyclic since each edge goes back in time and the loops are not allowed.

Figure 2.2 shows a piece of citation network which consists of papers belonging to one community and published in different years. For each paper i, we consider its reference list length R_{0i} (the number of outgoing edges, out-degree in network parlance) and the number of citations K_i (the number of ingoing edges, in-degree in network parlance). Although references and citations represent two sides of the same coin, statistical distributions of R_{0i} and K_i are very different even for the papers belonging to the same community. This is schematically illustrated in Fig. 2.2: for each node, the number of outgoing edges (references) is the same for

© The Author(s), under exclusive license to Springer Nature Switzerland AG 2019
M. Golosovsky, *Citation Analysis and Dynamics of Citation Networks*,
SpringerBriefs in Complexity, https://doi.org/10.1007/978-3-030-28169-4_2

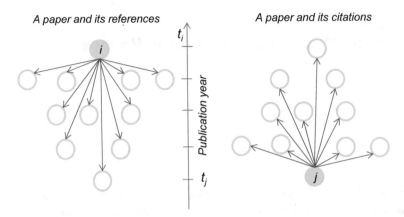

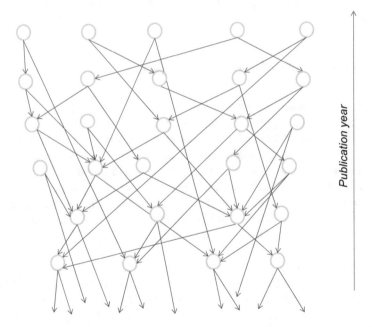

Fig. 2.1 The left panel shows a basic unit of a citation network—a paper and its references. The filled circle indicates a paper i published in year t_i, the open circles show its references published in years $t_i - 1, t_i - 2, \ldots$. The red lines connect the new paper with its references. Direction of the arrow shows who cites whom. The right panel shows a paper and its citations. The filled circle shows a paper j published in year t_j, the open circles show its citing papers published in years $t_j + 1, t_j + 2, \ldots$. In contrast to the left panel, where each new node appears together with all its outgoing edges, the number of ingoing edges for each node in the right panel permanently grows

Fig. 2.2 Citation network consisting of papers belonging to one community. The circles are papers, the edges are citations, direction of the arrows indicates who cites whom. Note the difference between the number of incoming and outgoing edges (in-degree and out-degree in network parlance). While the number of outgoing edges is narrowly distributed, the number of ingoing edges has broad distribution. In particular, each node has exactly 2 outgoing edges (references) and from 0 to 5 ingoing edges (citations). Citation network is not stationary and both the number of nodes and the number of edges increase with time

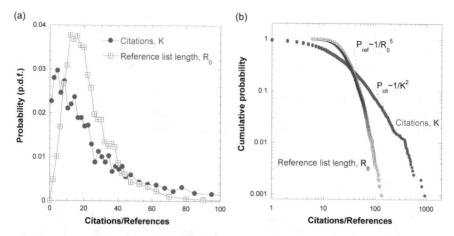

Fig. 2.3 Statistical distribution of the reference list lengths, $p(R_{0i})$, and of the long-time limit of citations, $p(K_i)$, for the same set of papers (all 2078 Physical Review B papers published in 1984). (**a**) Probability density function. While both distributions have almost the same mean, $p(R_{0i})$ distribution is a narrow bell-shaped distribution, while $p(K_i)$-distribution is very wide and has a fat tail. (**b**) Cumulative distributions, $P(X) = \int_X^\infty p(x)dx$. Both distributions have a power-law tail, $P(X) \sim 1/X^{\nu-1}$, but the exponent of the citation distribution is relatively small, $\nu = 3$, hence it belongs to the class of the fat-tailed distributions; while the exponent of the $P(R_0)$-distribution is big, $\nu = 6$, hence it is a regular distribution. Adapted with permission from Golosovsky and Solomon [72]. Copyright (2017) by the American Physical Society. https://journals.aps.org/pre/abstract/10.1103/PhysRevE.95.012324

all papers, $R_{0i} = 2$, while the number of incoming edges (citations) varies in broad limits, from $K_i = 0$ to $K_i = 5$.

Figure 2.3 shows statistical distributions of the reference list length $p(R_{0i})$ and citations $p(K_i)$ for the same set of papers. While $p(R_{0i})$ distribution is a relatively narrow bell-shaped curve, citation distribution is extremely broad. This is quite common for complex networks, in particular, the WWW exhibits a similar asymmetry between the in- and out-degree distributions [21].

What is the source of asymmetry between statistical distributions of references and citations? To our opinion, it derives from the fact that the references of a paper are compiled by one author while citations of a paper derive from many authors. Indeed, although the journal style does not standardize the reference list length, the authors make their best to comply with what is accepted in their research field. This means that there is a feedback mechanism that forces the authors to adhere to some average reference list length and this feedback results in a relatively narrow $p(R_{0i})$-distribution. Indeed, Ref. [171] measured the reference list length distribution in several Physics journals and showed that for the journals with a page limit the $p(R_{0i})$-distribution is narrow while for the journals without page limit this distribution is broader but not as wide as citation distribution.

However, when we consider citation dynamics of a paper—the decision on whether to cite it or not comes from many uncoordinated authors. There is no feedback mechanism regulating the number of citations of a paper and this is the source of enormous variability in citation dynamics of individual papers.

In summary, the reference list composition results from the decisions of an author and it is determined by psychology, fashion, and style. Hence, its analysis should be better performed using methods of the sociology of science. However, citations are the result of decisions of many independent authors or groups of authors, hence statistics of citations depends on the structure of citation network and it is more amenable to analysis by the physical and statistical methods.

2.2   Expanding Science

The science is constantly expanding and this is reflected in the nearly exponential growth of publications, authors, and journals. In particular, Sinatra et al. [155] found ~2.5% annual growth of the number of Physics papers during the last century. This growth is not homogeneous—in the period of 1950–1970 the annual growth was explosive and amounted to 5.35%, while since 1970 it became steady and averaged 1.7%. This is consistent with findings of Ref. [53] which reported 1.43% average growth rate in the period 1900–2017. The findings of Ref. [18] (1.26% annual growth of all scientific publications in the period 1980–2010) also corroborate this steady rate.

We used the Clarivate Web of Science database to measure $N(t)$, the numbers of the Physics, Mathematics, and Economics papers published annually during the period 1980–2015. Figure 2.4 shows that, on average, these three fields underwent

Fig. 2.4 Annual number of the Physics, Mathematics, and Economics publications. The growth exponents for these three disciplines are close but not identical. The growth is not uniform, each discipline exhibits periods of fast and slow growth

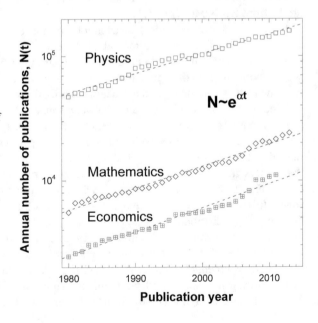

exponential growth, $N(t) \propto e^{\alpha t}$ with $\alpha = 0.036$–0.044, corresponding to 1.5–1.9% annual growth. Interestingly, the exponents for all three disciplines are close, as if their growth has been synchronized. This synchronization can be related to the interdisciplinary research and cross-fertilization between the disciplines. Indeed, community structure detection [52, 155] shows that natural sciences strongly overlap. This overlap is the most probable reason of synchronization, similar to what was found for the economic growth of countries [157]. Although the growth rate of all three mature fields is more or less the same, the growth rate of new fields, that are not strongly coupled to old body of science, can be higher, in particular, the rate of growth of computer science publications in 1970–2010 was 3.26% [166] and this is twice as fast as compared to the growth rate of Physics during the same period.

The reference list length also grows with time albeit more slowly [51, 89, 106]. In particular, Ref. [89] claimed logarithmic time dependence while Ref. [51] claimed a very slow growth before 2000 and subsequent acceleration following the advent of open access and electronic format journals that have no page limit. Figure 2.5 shows our measurements of $R_0(t)$, the average length of the reference list of the Physical Review B papers published in 1996–2013. We find a weak exponential dependence which is consistent with previous observations.

Not only citation network constantly grows, it densifies with time. In particular, Lescovec et al. [97] noticed that the number of citations grows faster than the number of papers. To illustrate this, we denote by $N(t)$ and $C(t)$, correspond-

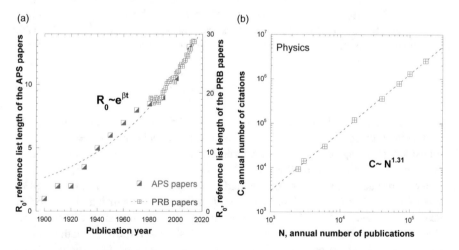

Fig. 2.5 (a) Time dependence of the average reference list length of the Physical Review papers. The violet squares show measurements of Ref. [89] for the corpus of Physical Review papers (only PR to PR references are counted), the blue squares show our measurements for the Physical Review B papers. Continuous line shows exponential approximation $R_0 \propto e^{\beta t}$ where $\beta = 0.014$. (b) Total number of annual citations in dependence of the annual number of Physics publications (from the data of Ref. [155]). The number of citations C grows faster than the number of papers N, namely, citation network densifies with time

ingly, the number of new papers that appear at time t and the number of new references/citations. Obviously, $C(t) = N(t)R_0(t)$. Since both $N(t)$ and $C(t)$ grow nearly exponentially, $N(t) \propto e^{\alpha t}$ and $C(t) = N(t)R_0(t) \propto e^{(\alpha+\beta)t}$, then $C \propto N^{1+\frac{\beta}{\alpha}}$. For Physics papers, the densification exponent is $1 + \frac{\beta}{\alpha} \approx 1.31$ and this number is close to what has been observed in many other complex networks [97]. For a moment, it is not clear whether this densification leads to some phase transition or not, because most documented complex networks are still too young and too far from such transition.

2.3 Temporal Aspect of the Citation Network

2.3.1 Age Distribution of References

We consider now the temporal structure of citation networks. We focus on some paper i published in year t_i and denote by $R_i(t_i, t_i - t)$ the number of its references that were published in year $t_i - t$. The function $R_i(t_i, t_i - t)$ varies too strong from paper to paper, therefore, we introduce another function, $R(t_i, t_i - t) = \overline{R_i(t_i, t_i - t)}$, which is the average number of references in the reference list of papers that belong to one discipline and that were published in year t_i. We name this function the age distribution of references.

Obviously, $\int_0^\infty R_i(t_i, t_i - t)dt = R_{0i}$, the length of the reference list of paper i. We denote by $R_0(t_i) = \overline{R_{0i}}$ the average reference list length for all papers belonging to one discipline and that were published in year t_i. Following Refs. [37, 59, 63, 120, 143, 146], we consider the reduced age distribution of references,

$$r(t_i, t) = \frac{R(t_i, t_i - t)}{R_0(t_i)}. \tag{2.1}$$

The rationale behind the introduction of $r(t_i, t)$ is that, as it is illustrated in Fig. 2.6, it is almost independent of the publication year, namely, $r(t_i, t) = r(t)$. This robustness of the age distribution of references, which is also called retrospective or synchronous citation distribution [63, 120], distribution of the ages of cited papers, or recent impact method, was documented by many researchers [57, 59, 166, 183]. The invariant distribution of references with respect to semantics and position in text was noticed by [13]. Recent comprehensive measurements of Sinatra et al. [155] for Physics papers corroborated these conclusions and showed that the secular dependence of $r(t_i, t)$ on the publication year t_i is very weak. In particular, since $t_i = 1980$ the difference between $r(t_i, t)$ for different t_i and the same t appears mostly at $t < 3$. This is probably related to the fact that the publishing has become easier after electronic correspondence with the editor replaced the surface mail.

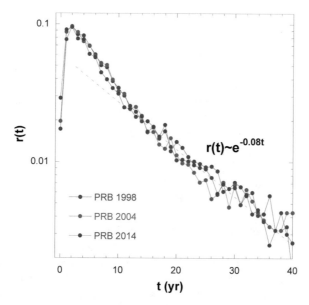

Fig. 2.6 $r(t)$, reduced age distribution of references, namely, a fraction of references published in year $t_i - t$ that appear in the reference list of a paper published in year t_i. Red, blue, and brown circles stay for three sets of research papers published in July issues of the Physical Review B in 1998, 2004, and 2014, correspondingly. Similar to previous studies [59, 143], we find that $r(t)$ dependences for all publication years collapse onto a single curve (with the possible exception of $t = 0$). The red dashed line shows exponential fit to the tail, $r(t) \sim e^{-0.08t}$ for $t > 10$. Adapted with permission from Golosovsky and Solomon [72]. Copyright (2017) by the American Physical Society. https://journals.aps.org/pre/abstract/10.1103/PhysRevE.95.012324

Figure 2.6 shows our measurements of $r(t_i, t)$ for the Physical Review B papers. The almost universal behavior of $r(t_i, t)$ for different t_i, as documented in this figure, together with the narrow distribution of the reference list lengths R_{0i} (Fig. 2.3), imply that the age composition of the reference lists of all papers in one discipline differ only in scale, in other words,

$$R_i(t_i, t_i - t) \approx r(t)R_{0i}. \tag{2.2}$$

2.3.2   Age Distribution of Citations

In order to extend the above formalism to citations, we consider a paper j published in year t_j and denote its annual citation rate as $k_j(t_j, t_j + t)$. This function indicates the number of citations that the paper j garnered in the year $t_j + t$. The corresponding cumulative number of citations is $K_j(t) = \int_0^t k_j(t_j, t_j + \tau)d\tau$. We also introduce the mean citation rate, $m(t_j, t_j + t) = \overline{k_j}(t_j, t_j + t)$, where the averaging is performed over all papers in one field which were published in year t_j.

The function $m(t_j, t_j + t)$, which is known as diachronous or prospective citation distribution [63, 120], is the analog of the $R(t_i, t_i - t)$ function for references. However, there are significant differences between the two. First, unlike $R(t_i, t_i - t)$, the function $m(t_j, t_j + t)$ has pronounced dependence on the publication year and varies from discipline to discipline [166, 183]. Secondly, while the cumulative number of references, $\int_o^t R(t_i, t_i - \tau)d\tau$, converges to $R_0(t_i)$ for $t \rightarrow \infty$, the cumulative number of citations, $M(t_j, t_j + t) = \int_o^t m(t_j, t_j + \tau)d\tau$, may diverge for $t \rightarrow \infty$.

Statistical distributions of R_{0i} and $K_j(t_j, t_j+t)$ for $t \rightarrow \infty$ are also very different (Fig. 2.3). Narrow R_{0i} distribution indicates that the age structure of the reference lists of individual papers is well represented by the mean, $R(t_i, t_i-t) \approx \overline{R_i}(t_i, t_i-t)$. On the contrary, the broad $K_j(t_j, t_j+t)$ distribution indicates that citation dynamics of individual papers varies in broad limits and can be qualitatively different from the mean. Indeed, Fig. 2.7 shows that different cohorts of Physics papers, all published in the same year, have strongly different citation rates. While citation rate of most papers increases after publication, goes through the maximum, and then gradually decays; citation rate of the highly-cited papers does not necessarily decay with time

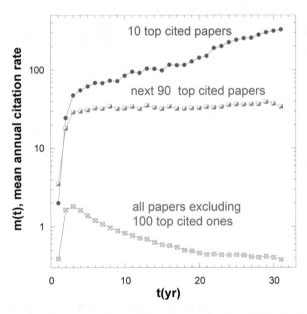

Fig. 2.7 Citation rate of several cohorts of Physics papers, all published in the same year, $t_j = 1984$. While citation rate of the average paper passes through a maximum and then decreases, citation rate of the highly-cited papers does not decay and they can be cited indefinitely long. Adapted with permission from Golosovsky [67]. Copyright (2017) by the American Physical Society. https://journals.aps.org/pre/abstract/10.1103/PhysRevE.96.032306

and can even increase. The implication that follows: unlike references, for which $R(t_i, t_i - t)$ truly represents the age composition of the reference list of a typical paper published in year t_i, the function $m(t_j, t_j + t)$ hardly represents citation dynamic of a paper published in year t_j.

2.3.3 Reference-Citation Duality

Despite considerable difference between statistical distributions of references and citations, they represent the two sides of the same coin. Hence, there is a certain relation between citation dynamics of papers, on the one hand, and the age distribution of references, on another hand. In what follows, we explore mathematical consequences of this relation.

We come back to the complex network representation of the papers and their references and citations. Figure 2.8 focuses on two sets of papers published in years t_a and t_b. We denote by $N(t_a)$ and $N(t_b)$ the number of papers in each set. We also introduce $m(t_a, t_b)$, the mean number of citations garnered in year t_b by a paper published in t_a; and $R(t_b, t_a)$, the average number of references published in year t_a that appear in the reference list of a paper published in year t_b. We assume that all citing papers belong to the same research field and neglect interdisciplinary

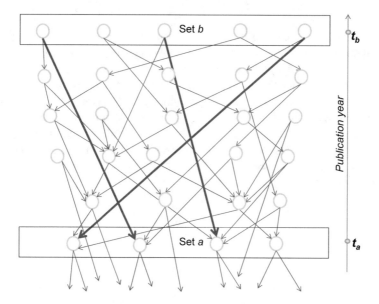

Fig. 2.8 Schematic representation of the reference-citation duality. Consider two sets of papers published in years t_a and t_b. The bold lines show the links between these two sets. From the perspective of set a, these links are citations. From the perspective of set b, these links are references

papers, books, and other references/citations which are not research papers. Under this assumption, the number of papers that cite the set a and that were published in year t_b is equal to the number of references published in year t_a that appear in the reference lists of the papers published in t_b, namely,

$$N(t_a)m(t_a, t_b) = N(t_b)R(t_b, t_a).$$ (2.3)

We introduce $t = t_b - t_a$, in such a way that $t_b = t_a + t$. In Eq. 2.3 we replace $R(t_b, t_a)$ by $r(t)R_0(t_b)$ where $R_0(t_b)$ is the average length of the reference list of the papers published in t_b, and $r(t)$ is the reduced age distribution of references (Eq. 2.2). Since the number of publications grows exponentially, $N(t_b) = N(t_a)e^{\alpha t}$ (Fig. 2.4), and the reference list length also grows exponentially, $R_0(t_b) = R_0(t_a)e^{\beta t}$ (Fig. 2.5a), then Eq. 2.3 reduces to

$$m(t_a, t_a + t) = R(t_a, t_a - t)e^{(\alpha+\beta)t},$$ (2.4)

where we used the relation $R(t_a, t_a - t) = R_0(t_a)r(t)$. In what follows, we consider the mean number of citations $m(t_a, t_a + t)$ as a function of t where publication year t_a is a parameter. We also consider the age distribution of references, $R(t_b, t_b - t)$, as a function of t where publication year t_b is a parameter. Equation 2.4 yields the relation between these two functions,

$$m(t_a, t_a + t) = R(t_b, t_b - t)e^{(\alpha+\beta)t}e^{\beta(t_a-t_b)},$$ (2.5)

where we used the equality, $R(t_b, t_b - t) = R(t_a, t_a - t) = e^{\beta(t_a-t_b)}$. Equation 2.5 captures the reference-citation duality. It relates synchronous (retrospective) and diachronous (prospective) citation distributions [37, 59, 63, 120, 143, 146, 183].

Figure 2.9 illustrates this duality. It shows that the functions, $m(t_a, t_a + t)$ and $R(t_b, t_b - t)$ exhibit almost a mirror-like symmetry. However, these functions are not identical due to the factor $e^{(\alpha+\beta)t}$. This factor is also responsible for a subtle qualitative difference between these two functions. Indeed, the integral $\int_0^t R(t_b, t_b - \tau)d\tau$ converges to $R_0(t_b)$ as $t \to \infty$. However, the function $m(t_a, t_a + t)$ decays slower due to exponential factor $e^{(\alpha+\beta)t}$, in such a way that the integral $\int_0^t m(t_a, t_a + \tau)d\tau$ can diverge as $t \to \infty$.

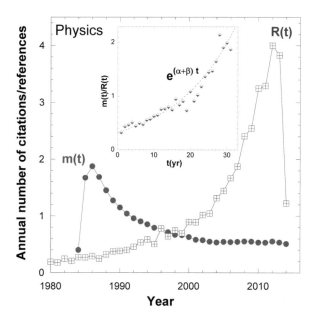

Fig. 2.9 The reference-citation duality. $m(t)$ is the mean annual number of citations of 48,168 Physics papers published in $t_a = 1984$. $R(t)$ is the age composition of the reference list of the Physical Review B papers published in $t_b = 2014$ (the data are from Fig. 2.6). Both dependences are qualitatively similar and display almost mirror-like symmetry. The inset shows the ratio of $m(t)$ to $R(t)$. The dashed line depicts exponential dependence $e^{(\alpha+\beta)t}$ suggested by Eq. 2.5 with $\alpha = 0.037$ yr^{-1} and $\beta = 0.014$ yr^{-1}. Adapted with permission from Golosovsky and Solomon [72]. Copyright (2017) by the American Physical Society. https://journals.aps.org/pre/abstract/10.1103/PhysRevE.95.012324

Chapter 3
Stochastic Modeling of References and Citations

Abstract We discuss here how the authors compose reference lists of their papers. We assume that the author chooses his references basing on the copying/recursive search algorithm. We put this algorithm at the core of the quantitative model accounting for the age composition of the reference lists of papers. The model contains two empirical functions: the aging function and the obsolescence function, and one empirical parameter—paper's fitness. At the next step, we calibrate this model in measurements with Physics papers, namely, we determine the aging function, obsolescence function, and paper's fitness averaged over the groups of similar papers. At the next step, we extend our approach to individual papers and formulate a probabilistic model of their citation dynamics. This model represents citation dynamic of a paper as a self-exciting stochastic process (Hawkes process). We demonstrate that the mean-field approximation of this probabilistic model yields the age distribution of references, as expected.

Keywords Citation models · Recursive search · Stochastic model · Hawkes process · Self-exciting process · Bass model

3.1 Copying/Recursive Search Model of the Citation Process

3.1.1 Scenario: The Author's Strategy to Compose the Reference List of a Paper

We believe that the composition of the reference lists of scientific papers is the clue to citation analysis. While citation dynamic of a paper is determined by several factors such as popularity of the research field, journal impact factor, preferences and tastes of citing authors, etc.; the reference list of a paper derives from only one source: decisions of an author (research team) who chooses the references basing on their content and age. We focus here on the age of the references and do not consider their content, although it can be very important [108].

© The Author(s), under exclusive license to Springer Nature Switzerland AG 2019   19
M. Golosovsky, *Citation Analysis and Dynamics of Citation Networks*,
SpringerBriefs in Complexity, https://doi.org/10.1007/978-3-030-28169-4_3

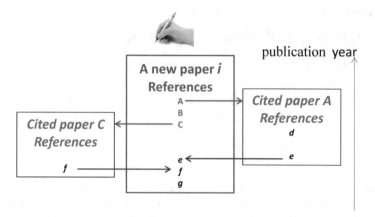

Fig. 3.1 Cartoon scenario of the referencing process which underlies our model. Consider a paper *i* and its list of references *A, B, C...e, f, g* arranged in descending chronological order. The author of *i* finds some references (*A, B, C*) and copies others (*e* and *f*)) from the reference lists of already selected papers (*A* and *C*, correspondingly.) We name the papers *A, B, C* direct references of *i*, while the papers *e* and *f* are indirect references

Our initial goal is to measure and to model the age composition of the reference lists of papers. To this end, we distinguish between direct and indirect references, as it is illustrated in Fig. 3.1. We define direct references as those papers that are not cited by any other reference of the parent paper. We define indirect references as the papers that are cited by one or several preselected references.

What is the source of indirect references? If the author cites some old seminal papers, then his most recent references will probably cite them as well. In our parlance, these old papers are indirect references. On another hand, indirect references may result from copying. Indeed, consider an author who writes a research paper. He reads scientific journals or media articles, searches databases, finds relevant papers and includes some of them in his reference list. These are direct references [89, 154, 170]. At the next step, the author studies the reference lists of these preselected papers, picks up relevant references, reads them, and adds some of them into his reference list. These are indirect references. Then the author studies the reference lists of the newly added papers, copies some references, and continues recursively. Our definition of indirect references includes all those papers that appear in the reference lists of the already selected papers, irrespective of the reason of such coincidence (copying or chance). In what follows we analyze the age composition of the reference list of papers generated by this copying/recursive search mechanism. Our analysis is based on the causality principle which requires the indirect references to be older than their preselected progenitors (Fig. 3.2).

The cartoon scheme of the referencing process is as follows. Consider a new paper that belongs to a certain community. We assume that the author of this paper can include into its reference list any older paper belonging to the same community. The choice of this paper can occur through two independent routes: the direct

A new paper

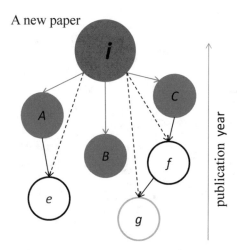

Fig. 3.2 Direct and indirect references. Consider a parent paper i and its references $A, B, C\ldots e,$ f, g arranged in descending chronological order. The papers A, B, C are not cited by any other papers in the reference list of i and we name them direct references. The papers e and f appear in the references lists of the already selected papers A and C, correspondingly, and we name them indirect references. The paper g appears in the reference list of the preselected paper f and it is also indirect reference. Each indirect reference closes a triangle in which the parent paper i is one of the vertices. The solid and dashed lines connect, correspondingly, the direct and indirect references with the parent paper i

and the indirect one. Namely, the author of the new paper i searches through the community, to which this paper belongs, and selects there some papers A, B, C as references. Then he analyzes the reference lists of the already selected papers, picks up some references, such as e, f, and includes them in his reference list as well. The process continues recursively, namely, at the next step the author analyzes the reference lists of the newly chosen papers f, picks up there another references g, includes them in his reference list, and so on. The papers A, B, C are direct references, while the papers e, f, g are indirect references (Fig. 3.2).

3.1.2 Recursive Search Model: Mathematical Formalism

To arrange these considerations into a formal mathematical framework, we consider a new paper i published in year t_i and an older paper j published in year t_j. The new paper i can choose the paper j directly, basing on its fitness and age. We assume that the probability of such direct citation is

$$\Pi_{ij}^{dir} = \eta_j \frac{R_{0i}}{N(t_j)} A(t_i, t_j), \tag{3.1}$$

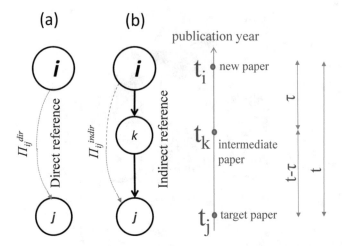

Fig. 3.3 Conceptual scheme of the citation process based on the recursive search algorithm. (**a**) Direct reference. A new paper i finds an older paper j through the fitness-based search among all old papers and includes it into its reference list with probability Π_{ij}^{dir} which depends on the fitness of the paper j and its publication year t_j (Eq. 3.1). (**b**) Indirect reference. The paper i finds an older paper j by analyzing the reference list of an already selected paper k. The paper i includes j into its reference list with probability Π_{ij}^{indir} which is determined by t_k, the publication year of the intermediate paper k, and by the length of its reference list R_{0k} (Eq. 3.2)

where η_j is the paper's fitness which characterizes the appeal that the paper j makes to the paper i, R_{0i} is the length of the reference list of i, $N(t_j)$ is the number of papers published in year t_j, and $A(t_i, t_j)$ is the aging function. (Later on, we will show that this function is the same for all papers and depends only on $t = t_i - t_j$, the age of the paper j with respect to the paper i.)

The logic behind Eq. 3.1 is the separation between the qualitative and temporal aspects affecting the author's choice of papers for the reference list of his new paper. Indeed, the appeal that the paper j makes to the author of the paper i depends first and foremost on its relevance which is determined by the overlap of their contents, the "quality" of paper j, and its age with respect to i. Since we assumed that all papers belong to the same community, we do not consider here the overlap of contents or similarity as it was done in Ref. [128]. The "quality" of the paper j is gauged by fitness η_j while the relative age is gauged by the function $A(t_i, t_j)$. Also, the probability of choosing the paper j increases with R_{0i}, the length of the reference list of the paper i (the longer the list, the higher the probability of choosing additional references). On another hand, the probability of choosing the paper j inversely depends on $N(t_j)$, the number of papers published in year t_j, since the author of i has to choose among $N(t_j)$ similar papers. These considerations justify Eq. 3.1 (see also [14, 29]).

Alternatively, the new paper i can pick up some old paper j from the reference list of a previously selected paper k. Among the reasons why the author of i picks up j, the age of the intermediate paper k is surely important. We can't know other reasons favoring the choice of j, so we make a simple assumption that after i has included k in its reference list, it can pick up any of its references with equal probability. We denote this probability by Θ_{ikj} and assume that it takes the following form:

$$\Theta_{ikj} = \frac{T(t_i, t_k)}{R_{0k}}, \tag{3.2}$$

where $T(t_i, t_k)$ is the obsolescence function which captures the fact that the probability of picking up an indirect reference from a recent paper is higher than of picking it from an old paper. R_{0k}, the length of the reference list of the intermediate paper k, which appears in the denominator of Eq. 3.2, captures our assumption that any reference of k can be picked up with equal probability. Our initial assumption is that $T(t_i, t_k)$ depends only on t_i and t_k. Later on we will revise this assumption.

We consider now the total probability of the indirect citation of the paper j by the paper i and denote it by Π_{ij}^{indir}. To write for it a compact expression, we switch to the network formalism, namely, we consider the complex network where papers are nodes and citations are edges. This network is directed and acyclic. To characterize its connectivity, we introduce the adjacency matrix, a_{mn}, which is defined as follows: $a_{mn} = 1$ if m cites n, otherwise $a_{mn} = 0$. Our initial assumption (to be revised later) is that Π_{ij}^{indir} is the sum over all intermediate papers k that contain paper j in their reference lists, namely, $\Pi_{ij}^{indir} = \sum_{k \in (\text{ref. list of } i)} \Theta_{ikj}$. Using adjacency matrix, we recast this expression as a sum over all papers l,

$$\Pi_{ij}^{indir} = \sum_l a_{il} a_{lj} \Theta_{ilj}. \tag{3.3}$$

We stress here that Eq. 3.3 is our starting assumption which we make in order to develop the model and to compare it to measurements. Later on, we will present the measurements which reveal that, in fact, the probabilities Θ_{ikj} do not sum up but interfere. This will prompt us to replace Eq. 3.3 by something like $\Pi_{ij}^{indir} \approx \left[\sum_k a_{ik} a_{kj} \Theta_{ikj}^{\frac{1}{2}} \right]^2$. We relegate these issues to Chap. 4.

In what follows, we assume that the total probability of a new paper i to cite an older paper j is the sum over direct and indirect routes, namely, $\Pi_{ij} = \Pi_{ij}^{dir} + \Pi_{ij}^{indir}$. Under this assumption, Eqs. 3.1–3.3 yield

$$\Pi_{ij} = (1 - a_{ij}) \left[\eta_j \frac{R_{0i}}{N(t_j)} A(t_i, t_j) + \sum_k a_{ik} a_{kj} \frac{T(t_i, t_k)}{R_{0k}} \right], \tag{3.4}$$

where the factor $(1 - a_{ij})$ takes into account that paper i can cite paper j only once. The normalization condition for Π_{ij} is

$$\sum_j \Pi_{ij} = R_{0i}, \qquad (3.5)$$

where the sum is over all papers. This condition imposes the scale of fitnesses η_j which will be discussed later.

3.1.3 Recursive Search Algorithm

Our model assumes that the author of a new paper i composes his reference list according to the following procedure:

1. The target length of the reference list is set to R_{0i} and the running length of this list is set to zero, $\tilde{R}_{0i} = 0$.
2. For each paper l, the probability of being cited, Π_{il}, is calculated according to Eq. 3.4.
3. Given set of probabilities Π_{il}, one paper j is chosen.
4. The adjacency matrix element for this paper is updated, $a_{ij} = 1$.
5. The length of the running reference list is updated, $\tilde{R}_{0i} = \tilde{R}_{0i} + 1$.
6. If $\tilde{R}_{0i} < R_{0i}$ then go to (2), else stop.

This algorithm captures our initial model and after model calibration it will be updated. It mimics the process by which most authors compose their reference lists. The corresponding model (Eq. 3.4) includes two empirical functions A and T which shall be found from measurements. This procedure is named model calibration and this is what we discuss further in this chapter and in Chap. 4. In addition, our model is based on several assumptions which require validation and this is the subject of Chap. 5. It should be also noted, that our model focuses on the temporal aspect of citation process and leaves aside other aspects, such as similarity between the citing and cited papers. This last assumption is justified so far as we deal with the papers belonging to one community.

3.2 Modeling Age Distribution of References in the Reference Lists of Scientific Papers

3.2.1 A Mean-Field Model for References

We consider a paper i published in year t_i and introduce $R_i(t_i, t_j)dt$, the age distribution of references. This function counts the number of references in the reference list of i which were published in the time window $(t_j, t_j + dt)$. It satisfies the normalization condition $\int_0^{t_i} R_i(t_i, t_j)dt_j = R_{0i}$, where R_{0i} is the length of reference list of i. Using adjacency matrix, this function can be recast as $R_i(t_i, t_j)dt = \sum_{j \in (t_j, t_j + dt)} a_{ij}$. On another hand, $R_i(t_i, t_j)dt = \sum_{j \in (t_j, t_j + dt)} \Pi_{ij}$. Then, Eq. 3.4 yields

$$R_i(t_i, t_j)dt = \sum_{j \in (t_j, t_j + dt)} \left[\eta_j \frac{R_{0i}}{N(t_j)} A(t_i, t_j) + \sum_k a_{ik} a_{kj} \frac{T(t_i, t_k)}{R_{0k}} \right], \quad (3.6)$$

where, for simplicity, we dropped the factor $(1 - a_{ij})$. We perform summation by j, drop dt, and come to $R_i(t_i, t_j) = \overline{\eta}_j R_{0i} A(t_i, t_j) + \sum_k a_{ik} T(t_i, t_k) \frac{R(t_k, t_j)}{R_{0k}}$. Here, $\overline{\eta}_j = \frac{1}{N(t_j)} \sum_{j \in (t_j, t_j + dt)} \eta_j$ is the average fitness of papers published in year t_j. At the next step, we shift to continuous approximation, replace the sum over k by the integral, and come to

$$R_i(t_i, t_j) = \overline{\eta}_j R_{0i} A(t_i, t_j) + \int_{t_j}^{t_i} \frac{R_k(t_k, t_j)}{R_{0k}} T(t_i, t_k) R_i(t_i, t_k) dt_k. \quad (3.7)$$

This equation expresses $R_i(t_i, t_j)$ through empirical functions $A(t_i, t_j)$ and $T(t_i, t_j)$. In what follows we obtain useful conclusions with respect to these functions by considering the reduced age distribution of references, $r_i(t_i, t_j) = \frac{R_i(t_i, t_j)}{R_{0i}}$. To this end, we divide Eq. 3.7 by R_{0i}, note that $\frac{R_k(t_k, t_j)}{R_{0k}} = r_k(t_k, t_j)$, and come to

$$r_i(t_i, t_j) = \overline{\eta}_j A(t_i, t_j) + \int_0^t \overline{r}_k(t_k, t_j) T(t_i, t_k) r_i(t_i, t_k) dt_k. \quad (3.8)$$

While $r_i(t_i, t_j)$ characterizes a single paper i, $\overline{r}_k(t_i, t_k)$ is an average over all references of i that were published in year t_k. In the derivation of Eq. 3.8 we assumed that $r_i(t_i, t_k)$ and $\overline{r}_k(t_k, t_j)$ are uncorrelated. We made this naive assumption only because the measurements of such correlation for references have been absent. Later on, we will revise this assumption.

A big advantage of $r_i(t_i, t_j)$ over $R_i(t_i, t_j)$ is that the former does not depend separately on t_i and t_j but is determined only by their difference $t = t_i - t_j$. We average Eq. 3.8 over all papers i which were published in year t_i, and come to

$$r(t) = \overline{\eta}_j \overline{A}(t_i, t_i - t) + \int_0^t r(\tau) \overline{T}(t_i, t_i - \tau) r(t - \tau) d\tau, \quad (3.9)$$

where $t = t_i - t_k$, $\tau = t_k - t_j$, $r(t) = r(t_i, t_i - t)$, and the correlations were neglected. $\overline{A}$ and $\overline{T}$ are, correspondingly, the aging and obsolescence functions averaged over all papers i published in year t_i. Since $r(t)$ does not depend explicitly on t_i, (see Fig. 2.6) we come to important conclusion that the functions $\overline{A}(t_i, t_i - t)$ and $\overline{T}(t_i, t_i - \tau)$ do not explicitly depend on t_i as well. Thus, these can be recast as $\overline{A}(t)$ and $\overline{T}(\tau)$, correspondingly. (In what follows, we demonstrate that $A(t)$ is nearly identical for all papers, namely, $\overline{A}(t) = A(t)$.)

The universality of $r(t)$ also means that $\overline{\eta}_j$, the average fitness of the papers published in year t_j, is independent on t_j, namely, $\overline{\eta}_j = \overline{\eta}$. To find $\overline{\eta}$, we consider the normalization condition, $\int_0^\infty r(\tau)d\tau = 1$. Accordingly, we integrate Eq. 3.9 over time and come to the following expression: $\overline{\eta} \int_0^\infty \overline{A}(\tau)d\tau + \int_0^\infty \overline{T}(\tau)r(\tau)d\tau = 1$. Next, we note that Eq. 3.9 does not contain the fitness $\overline{\eta}$ and aging function $\overline{A}(t)$ separately but depends only on their product. This allows us to normalize the aging function as follows, $\int_0^\infty \overline{A}(\tau)d\tau = 1$. Under such normalization, Eq. 3.9 yields

$$\overline{\eta} = 1 - \int_0^\infty \overline{T}(\tau)r(\tau)d\tau. \tag{3.10}$$

Finally, Eq. 3.9 reduces to

$$r(t) = r_{dir}(t) + r_{indir}(t) \tag{3.11}$$

where

$$r_{dir}(t) = \overline{\eta}\,\overline{A}(t), \tag{3.12}$$

$$r_{indir}(t) = \int_0^t r(\tau)\overline{T}(\tau)r(t-\tau)d\tau, \tag{3.13}$$

and $\overline{\eta}$ is given by Eq. 3.10.

Equations 3.11–3.13 include two empirical functions, $\overline{A}(t)$ and $\overline{T}(t)$. Once these functions are known, Eqs. 3.11 and 3.13 can be solved to find $r(t)$. In what follows, we perform the inverse procedure, namely, we measure the direct and indirect components of $r(t)$ and infer the functions $\overline{A}(t)$ and $\overline{T}(t)$ from these measurements. We name this a calibration procedure.

3.2.2 Model Calibration: Direct and Indirect References

Figure 3.4a shows measured functions $r(t)$, $r_{dir}(t)$ and $r_{indir}(t)$ for Physics papers. We observe that $r_{dir}(t)$ *sharply* increases during first couple of years after publication and then slowly decays, while $r_{indir}(t)$ *gradually* increases and then slowly decays, as expected.

How these findings are related to the previous knowledge? It shall be noted, that our model is loosely based on "the mathematical theory of citing" suggested by Simkin and Roychowdhury [154]. They were the first to construct the quantitative model of citation process based on copying mechanism. Although the model of Simkin and Roychowdhury is clever and insightful, it remained speculative since the

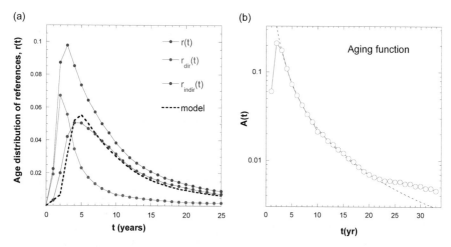

Fig. 3.4 (**a**) Time dependences of r_{dir}, r_{indir}, and $r = r_{dir} + r_{indir}$, the fractions of the direct, indirect, and total references in the reference list of an average Physical Review B paper. The data represent an average over 21 PRB papers published in 2014. The dashed line stays for model prediction based on Eq. 3.13 with exponential kernel $\overline{T}(\tau) = T_0 e^{-\gamma\tau}$, where $T_0 = 7.6$ and $\gamma = 1.2$ yr^{-1}. (**b**) Aging function for references as inferred from $r(t)$. It sharply increases during first couple of years after publication and then slowly decreases. The dashed line shows an empirical power-law fit, $A(t) = \frac{1.4}{(t+0.25)^{1.75}}$, where publication year corresponds to $t = 1$. Adapted with permission from Golosovsky and Solomon [72]. Copyright (2017) by the American Physical Society. https://journals.aps.org/pre/abstract/10.1103/PhysRevE.95.012324

supporting measurements were based on the anecdotal evidence inferred from the citation analysis of the highly-cited papers and not from the analysis of references. In particular, Ref. [154] assumed that $r_{dir}(t)$ decays very fast and effectively goes to zero in a couple of years, in such a way that the referencing process reduces to a ridicule: the author reads few recent papers and copies all other references from these recent papers. Our measurements reveal a much more realistic scenario of the referencing process: We find a relatively large proportion of direct references and a long tail of $r_{dir}(t)$, consistent with long decay of attention in social networks [30]. This means that an average author reads not only the recent but the old papers as well, and these non-copied papers make at least 30% of all references. Thus, in stark contrast to popular misconception that old papers are never read and become obsolete, we find that the authors do read old papers.

With respect to indirect references, Fig. 3.4 yields exponential kernel

$$\overline{T}(\tau) = T_0 e^{-\gamma t}, \tag{3.14}$$

where $T_0 = 7.6$ and $\gamma = 1.2$ yr^{-1}. Fast exponential decrease of $\overline{T}(\tau)$ suggests that if the references are copied—this is preferably done from the preselected references that were most recently published. The large proportion of indirect references in the reference list of papers conforms well with previous estimates of Refs. [42, 66,

154]. Quantitatively, consider a paper published in year t_0. From each preselected reference published in the same year, the author picks up $\sim$7.6 indirect references, from each 1-year-old preselected reference he picks up $\sim$2.3 references, from each 2-year-old preselected reference he picks $\sim$0.7 references, and so on.

3.3 Stochastic Model of Citation Dynamics of Individual Papers

To account for citation dynamics of individual papers, we reformulate our cartoon scenario (Fig. 3.1) in terms of citations. Namely, we stay in the framework of the same model but shift the focus from the citing paper to the cited paper.

Figure 3.5a shows a source paper j published in year t_j and its citing papers published in subsequent years. To proceed further, we reformulate our definitions of the direct and indirect references (Fig. 3.2) in terms of citations. Namely, a direct citation is the paper that cites paper j and does not cite any other paper that cites j; while an indirect citation is the paper that cites both j and one or more of its citing papers. For example, A, B ,C cite j and these are direct citations. The papers e, g, f cite, correspondingly, B, f, C that cite j, and these are indirect citations.

Figure 3.5b shows a source paper j and its citing papers from a different perspective—the focus is on the two generations of citing papers. The direct citations are those that belong only to the first-generation while indirect citations are those that belong both to the first- and to the second-generation citing papers.

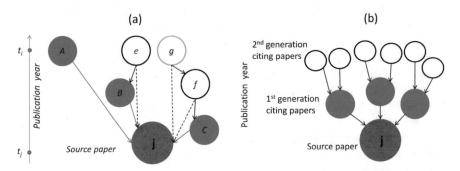

Fig. 3.5 (**a**) A fragment of citation network which contains a source paper j and its citing papers. A, B, C are direct citations since they cite j and do not cite any other paper citing j. The papers e, f cite B, C, correspondingly, and they are indirect citations. The paper g is also indirect citation since it cites f that cites j. The solid and dashed lines link the source paper with its direct and indirect citing papers, correspondingly. Each indirect citing paper closes a triangle in which the source paper j is one of the vertices. (**b**) Two generations of citing papers. The first-generation citing papers are those connected to the source paper j by a one-hop path. The second-generation citing papers are those connected to the source paper j by one or several two-hop paths

Due to permanent influx of new papers, the number of citations of each paper permanently grows. To quantify citation dynamics of a paper j that follows from our recursive search model, we introduce $\Delta K_j = k_j(t_j, t_i)dt$, the number of citations garnered by j in the time window $(t_i, t_i + dt)$. Here, $k_j(t_j, t_i)$ is the citation rate. We assume that ΔK_j is a random variable that follows a time-inhomogeneous stochastic point process. The probability of having ΔK_j citations in a short time interval dt is $\frac{\lambda_j^{\Delta K_j}}{\Delta K_j!}e^{-\lambda_j}$ where $\lambda_j dt$ is the paper-specific probabilistic citation rate which depends on time and on the previous citation history of a paper. Dependence on the previous history brings us into the realm of self-exciting or Hawkes processes. Our aim is to model its probabilistic rate λ_j.

The recursive search model of Sect. 3.1.3 yields $\lambda_j(t_j, t_i)dt = \sum_{i \in (t_i, t_i + dt)} \Pi_{ij}$. Using Eq. 3.4, we find

$$\lambda_j(t_j, t_i)dt = \sum_{i \in (t_i, t_i + dt)} \left[\eta_j \frac{R_{0i}}{N(t_j)} A(t_i - t_j) + \sum_k a_{ik} a_{kj} \frac{T(t_i - t_k)}{R_{0k}} \right], \quad (3.15)$$

where the factor $(1 - a_{ij})$ was dropped for simplicity.

Since we assumed that all papers belong to one community, the fitness of the paper j from the perspective of all papers i published in year t_i, is the same. Thus, $\sum_{i \in (t_i, t_i + dt)} \eta_j \frac{R_{0i}}{N(t_j)} A(t_i - t_j) = \eta_j R_0(t_i) \frac{N(t_i)}{N(t_j)} A(t_i - t_j)dt$ where $N(t_i)dt$ is the number of papers published in the time window $(t_i, t_i + dt)$ and $R_0(t_i)$ is their average reference list length. We also note that $\sum_{i \in (t_i, t_i + dt)} a_{ik} = k_k(t_k, t_i)dt$ where $k_k(t_k, t_i)$ is the citation rate of paper k in year t_i. After substitution of these expressions into Eq. 3.15, we obtain

$$\lambda(t_j, t_i) = \eta_j R_0(t_j) \frac{R_0(t_i)}{R_0(t_j)} \frac{N(t_i)}{N(t_j)} A(t_i - t_j) + \sum_k k_k(t_k, t_i) \frac{T(t_i - t_k)}{R_{0k}} a_{kj}, \quad (3.16)$$

where we replaced $R_0(t_i)$ by $R_0(t_j) \frac{R_0(t_i)}{R_0(t_j)}$ and dropped dt. Although the summation in the second addend is over all papers k, the factor a_{kj} singles out only those papers k which cite the paper j.

We note that both the number of publications and the reference list length grow exponentially, namely, $\frac{N(t_i)}{N(t_j)} = e^{\alpha(t_i - t_j)}$ and $\frac{R_0(t_i)}{R_0(t_j)} = e^{\beta(t_i - t_j)}$ (see Figs. 2.4 and 2.5). We substitute these expressions into Eq. 3.16, replace the sum by the integral, and come to

$$\lambda(t_j, t_i) = \eta_j R_0(t_j) A(t_i - t_j) e^{(\alpha + \beta)(t_i - t_j)} + \int_{t_j}^{t_i} k_k(t_k, t_i) \frac{T(t_i - t_k)}{R_0(t_k)} k_j(t_j, t_k)dt_k. \quad (3.17)$$

For simplicity, we assumed here that all papers k published in year t_k have the same reference list length, namely, we replaced R_{0k} by $R_0(t_k)$.

We note that $k_j(t_j, t_k)dt$ is the number of papers that cite j and that were published in the time window (t_k, t_k+dt), while $k_k(t_k, t_i)dt$ is the number of papers that cite the latter and that were published in the time window $(t_i, t_i + dt)$, in such a way that with respect to the paper j, the latter are second-generation citing papers. In the parlance of complex networks, $k_j(t_j, t_k)$ is the time-resolved degree (connectivity) of the paper j and $k_k(t_k, t_i)$ is its time-resolved nearest-neighbor connectivity. We denote the latter by $k_j^{nn}(t_k, t_i)$. If we consider an ensemble of papers j published in year t_j, then $k_j^{nn}(t_k, t_i)$ can be positively or negatively correlated to $k_j(t_j, t_k)$. In the language of complex network, the positive correlation is named assortativity and the negative correlation is named disassortativity. Temporarily, we assume that such correlations are absent. Later on, we will revise this assumption.

To compress notation, we introduce $t = t_i - t_j$ and $\tau = t_k - t_j$, in such a way that Eq. 3.16 transforms to

$$\lambda_j(t_j, t_j + t) = \eta_j R_0(t_j)\tilde{A}(t) + \int_0^t k_j^{nn}(t_j + t - \tau, t_j + t)\frac{T(\tau)}{R_0(t_k)}k_j(t_j, t_j + t - \tau)d\tau$$

(3.18)

where

$$\tilde{A}(t) = A(t)e^{(\alpha+\beta)t}$$

(3.19)

is the aging function for citations.

Since $k_j^{nn}(t_j + t - \tau, t_j + t)$ is already an average over a large ensemble of papers, we replace it by the mean number of citations, $m(t_j + t - \tau, t_j + t)$. Basing on Eqs. 2.4 and 2.5, we find $m(t_j + t - \tau, t_j + t) = m(t_j, t_j + \tau)e^{\beta(t-\tau)}$. We also note that $R_0(t_k) = R_0(t_j)e^{\beta(t-\tau)}$. We substitute these expressions into Eq. 3.18 and, using properties of convolution, we replace τ by $t - \tau$ in the integral. This yields

$$\lambda_j(t) = \eta_j R_0\tilde{A}(t) + \int_0^t m(t - \tau)\frac{T(t - \tau)}{R_0}k_j(\tau)d\tau,$$

(3.20)

where all functions pertain to the same publication year t_j, hence it was dropped from our notation. For example, $R_0(t_j)$ has been replaced by R_0.

The first addend in Eq. 3.20 represents the direct citations. Here, the fitness η_j is determined by the attributes of the paper j, while two other factors, $R_0(t_j)$ and $\tilde{A}(t)$, are the same for all papers published in the same year.

The second addend in Eq. 3.20 represents the indirect citations. It is determined by $k_j(\tau)$, the past citation rate of the paper j, and by the functions $m(t)$ and $\frac{T(t)}{R_0}$ which are common for all papers published in the same year.

As expected, Eq. 3.20 satisfies the duality principle (Eq. 2.5). To demonstrate this, we average Eq. 3.20 over all papers j published in year t_j and come to

$$m^{dir}(t_j, t_j + t) = \overline{\eta_j} R_0(t_j) A(t) e^{(\alpha+\beta)t} = R^{dir}(t_j, t_j - t) e^{(\alpha+\beta)t};$$ (3.21)

$$m^{indir}(t_j, t_j + t) = \int_0^t m(t - \tau) \frac{T(t - \tau)}{R_0(t_j)} m(\tau) d\tau = R^{indir}(t_j, t_j - t) e^{(\alpha+\beta)t}.$$ (3.22)

Thus, $m(t_j, t_j + t) = R(t_j, t_j + t) e^{(\alpha+\beta)t}$, in accordance with Eq. 2.4.

3.4 Continuous Approximation of the Model

To better understand Eq. 3.20, we analyze its continuous approximation, namely, we disregard stochasticity and replace λ_j, the latent citation rate of a paper j, by the actual citation rate k_j which we consider as a continuous variable. We develop this approximation with purely pedagogical purposes, it can not be used for quantitative estimates. We also switch to continuous time and replace the sum by the integral. In addition, we replace the kernel $m(t - \tau) \frac{T(t-\tau)}{R_0}$ by the exponent $q_j e^{\tilde{\gamma}(t-\tau)}$ where all time dependences are absorbed in $\tilde{\gamma}$ and all prefactors are absorbed in q_j. The reason for such drastic simplification is that $T(t)$ decays with time much stronger than $m(t)$. The continuous approximation of Eq. 3.20 then reads

$$k_j(t) = \eta_j R_0 \tilde{A}(t) + q_j \int_0^t k_j e^{-\tilde{\gamma}(t-\tau)} d\tau.$$ (3.23)

Dynamic behavior described by Eq. 3.23 results from the interplay between the positive feedback captured by the factor q_j and the obsolescence rate characterized by $\tilde{\gamma}$.

3.4.1 Relation to the Bass Model

For small $\tilde{\gamma}$, Eq. 3.23 reduces to the models of Refs. [108, 130, 133, 151],

$$k_j(t) \approx \eta_j R_0 \tilde{A}(t) + q_j K_j(t).$$ (3.24)

Equation 3.24 is nothing else but the Bass equation for diffusion of innovations in infinite market where citations correspond to adopters, direct citations correspond to innovators, and indirect citations correspond to imitators. The connection to the Bass model is not occasional since each paper is a new product whose penetration to the market of ideas is gauged by the number of citations [110].

For large $\tilde{\gamma}$, the main contribution to the integral in Eq. 3.23 comes from recent citations garnered in the short time interval $(t, t - 1/\tilde{\gamma})$. Having this in mind, we approximate $k_j(\tau)$ by $k_j(t) - (t - \tau)\frac{dk_j}{d\tau}|_t$, perform integration in the time window $(t, t - 1/\tilde{\gamma})$, and after some algebra arrive at

$$k_j(t) \approx \eta_j R_0 \tilde{A}(t) + \frac{q_j}{\tilde{\gamma}} k_j(t - 1/\tilde{\gamma}). \tag{3.25}$$

Equation 3.25 is the first-order autoregressive model of citation dynamic where time delay is $1/\tilde{\gamma}$ and $q_j/\tilde{\gamma}$ is the first-order autoregressive parameter [70]. In contrast to Eq. 3.24, which assigns equal weight to all past citations, Eq. 3.25 assigns more weight to recent citations. Since the measurements yield large $\tilde{\gamma}$, then Eq. 3.25 provides a more realistic description of citation dynamics of scientific publications than Eq. 3.24. Similar models were suggested by Refs. [154, 176]. Ref. [65] showed that dynamics of Facebook installation decisions is also biased toward recent rather than cumulative popularity.

3.4.2 Analytic Solution

We come back to Eq. 3.23. This is an integral Fredholm equation of the second kind. To solve it, we introduce a new variable $y_j = \int_0^t k_j e^{\tilde{\gamma}\tau} d\tau$, substitute it there, and obtain

$$\frac{dy_j}{dt} = \eta_j R_0 \tilde{A}(t) e^{\tilde{\gamma}t} + q_j y_j. \tag{3.26}$$

Equation 3.26 is easily integrated, yielding $y_j = \eta_j R_0 e^{q_j t} \int_0^t \tilde{A}(\tau) e^{(\tilde{\gamma} - q_j)\tau} d\tau$. We recall that $k_j = \frac{dy_j}{dt} e^{-\tilde{\gamma}t}$ and obtain

$$k_j(t) = \eta_j R_0 \left[\tilde{A}(t) + q_j \int_0^t \tilde{A}(\tau) e^{-(\tilde{\gamma} - q_j)(t - \tau)} d\tau \right]. \tag{3.27}$$

The first addend here corresponds to $k_j^{dir}(t)$ and the second addend corresponds to $k_j^{indir}(t)$. Integration of this equation yields cumulative number of citations,

$$K_j(t) = \int_0^t k_j(t_1) dt_1 = \eta_j R_0 B(t), \tag{3.28}$$

where

$$B(t) = \int_0^t \tilde{A}(t_1) dt_1 + q_j \int_0^t e^{-(\tilde{\gamma} - q_j)t_1} dt_1 \int_0^{t_1} \tilde{A}(\tau) e^{(\tilde{\gamma} - q_j)\tau} d\tau. \tag{3.29}$$

Equations 3.28 and 3.28 suggest that if $q_j = const$, then citation dynamics of the papers published in the same year are qualitatively similar and differ only by scale which is set by paper's fitness η_j, while citation dynamics for the papers in different disciplines differ by scale which is set by R_0. Citation trajectory of all papers is set by the factor $B(t)$ which slightly varies between different disciplines due to the different growth rates of the number of publications.

To analyze citation dynamic that follows from Eq. 3.27, we crudely approximate the aging function $\tilde{A}(t)$ by the exponent $A_0 e^{-\delta t}$, substitute it there, and come to

$$k_j^{dir}(t) = \eta_j R_0 A_0 e^{-\delta t}, \qquad (3.30)$$

$$k_j^{indir}(t) = \eta_j R_0 A_0 \frac{q_j}{\tilde{\gamma} - q_j - \delta} \left[e^{-\delta t} - e^{-(\tilde{\gamma} - q_j)t} \right]. \qquad (3.31)$$

Integration of Eq. 3.30 yields cumulative number of direct citations

$$K_j^{dir}(t) = \eta_j R_0 \frac{A_0}{\delta}(1 - e^{-\delta t}). \qquad (3.32)$$

The expression for $K_j^{indir}(t)$ is too cumbersome and we do not show it here.

As we will show later, $\tilde{\gamma} - q_j > 0$ and $\tilde{\gamma} - q_j \gg \delta$ for most papers. Equations 3.30 and 3.31 indicate that citation dynamics of a paper j is governed by the interplay of two exponentials, the slow one, $e^{-\delta t}$, and the fast one, $e^{-(\tilde{\gamma} - q_j)t}$. After a transient period with duration $\sim 1/(\tilde{\gamma} - q_j)$, indirect citations track direct citations, in such a way that the overall citation rate is governed by direct citations. Indirect citations appear as dividends of direct citations, each of the latter bringing $\sim \frac{q_j}{\tilde{\gamma} - q_j - \delta}$ indirect citations. In the long time limit, the total number of citations approaches saturation,

$$K_j^{\infty} = \eta_j R_0 \frac{A_0}{\delta} \frac{\tilde{\gamma}}{\tilde{\gamma} - q_j}. \qquad (3.33)$$

In the opposite case of $\tilde{\gamma} - q_j < 0$, Eqs. 3.30 and 3.31 predict the runaway behavior. In this case, citation rate of a paper accelerates in time and in the long-time limit grows exponentially,

$$k_j(t) \sim \eta_j R_0 A_0 \frac{q_j}{q_j - \tilde{\gamma} + \delta} e^{(q_j - \tilde{\gamma})t}. \qquad (3.34)$$

Chapter 4
Citation Dynamics of Individual Papers: Model Calibration

Abstract The model of citation dynamics developed in Chap. 3 is compared to measurements in order to determine empirical functions and parameters such as aging function, obsolescence function, and the paper's fitness. We found that the aging function is universal, namely, it is the same for all papers in one field published in the same year. However, the obsolescence function depends on the number of previous citations. This unexpected finding prompted us to focus more closely on the network aspect of citation dynamics and to consider not only the nearest neighbors of each paper in the citation network, but its next-nearest neighbors as well. The updated model takes into account the correlations between citation dynamics of a paper and its neighbors (network assortativity).

Keywords Complex networks · Assortativity · Degree-degree correlation · Nonlinear dynamics

4.1 Measurements

The model of citation dynamics developed in Chap. 3 contains several empirical functions and parameters. The goal of this chapter is to measure them. In addition, our model contains several assumptions. Verification of these assumptions is the subject of Chap. 5.

4.1.1 Methodology

The model developed in Chap. 3 states that citation dynamic of a paper j is determined by its fitness η_j, aging function $\tilde{A}(t)$, and obsolescence function $T(t)$. To organize our measurements, we started from an a priori assumption that the functions $\tilde{A}(t)$ and $T(t)$ are the same for all papers published in the same year and are equal to those found in Chap. 3 from the analysis of references. After performing the measurements, we will reassess this assumption.

© The Author(s), under exclusive license to Springer Nature Switzerland AG 2019
M. Golosovsky, *Citation Analysis and Dynamics of Citation Networks*,
SpringerBriefs in Complexity, https://doi.org/10.1007/978-3-030-28169-4_4

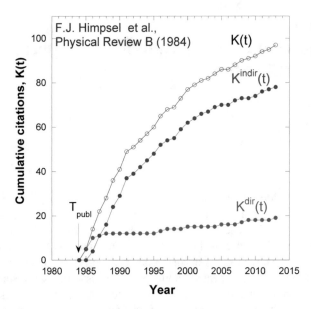

Fig. 4.1 Cumulative number of citations of a well-cited Physics paper Himpsel et al. [78]. The direct citations shoot up immediately after publication of the paper and after a couple of years their accumulation rate slowly decays. The indirect citations shoot up with 1–2 year delay with respect to direct citations, their rate peaks in another couple of years and then decays. Adapted with permission from Golosovsky and Solomon [72]. Copyright (2017) by the American Physical Society. https://journals.aps.org/pre/abstract/10.1103/PhysRevE.95.012324

The straightforward way of determining $\tilde{A}(t)$, $T(t)$, and η_j is to measure citation dynamics of individual papers, to discriminate between direct and indirect citations, and to fit their dynamics using Eq. 3.20 with η_j as a fitting parameter, and $\tilde{A}(t)$, $T(t)$ as fitting functions. Figure 4.1 illustrates such approach. It shows dynamics of direct and indirect citations for a Physics paper that garnered 100 citations in 25 years. For such a well-cited paper with a smooth citation trajectory, the fitting procedure should work. But for ordinary papers which garner 10–50 citations in the long time limit, the fitting procedure does not work well since citation trajectories of such papers are too noisy.

To minimize noise, we considered average citation dynamics of the groups of similar papers, namely, the papers that belong to the same research field, were published in the same year, and garnered the same number of citations K_j^∞ in the long-time limit. The index j denotes here a group of papers rather than a single paper. Our implicit assumption was that the papers in each group have the same fitness and their citation trajectories are governed by the same functions $\tilde{A}(t)$ and $T(t)$. Therefore, the mean citation rate of the papers in such a group is a good proxy to probabilistic citation rate of a paper with a given fitness, namely, $\overline{k_j}(t) \approx \lambda_j(t)$.

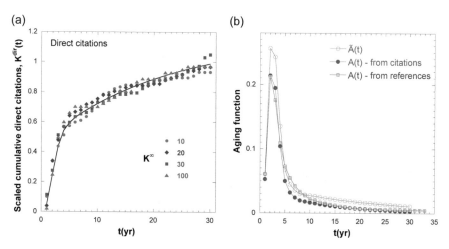

Fig. 4.2 (**a**) Scaled cumulative direct citations for 37 Physical Review B papers published in 1984. Each set of points represents the data averaged over a group of papers that garnered the same number of citations by the end of 2013, namely $K_j^\infty = 10, 20, 30,$ and 100. The data for each group was divided by an appropriate factor to achieve collapse of all $K_j^{dir}(t)$ trajectories onto one curve. These collapsed data were smoothed, producing the black continuous curve which we used to calculate the aging function $\tilde{A}(t)$. (**b**) The aging function for citations, $\tilde{A}(t)$, as found by differentiation of the black curve in (**a**). $A(t)$ is the aging function for references. It was calculated from $\tilde{A}(t)$ using Eq. 3.19 with $\alpha = 0.037$ yr^{-1} and $\beta = 0.014$ yr^{-1}. We also replot here $A(t)$ found in our studies of references (Fig. 3.4). The aging functions $A(t)$ found from the measurements of citations and references are almost the identical, as expected. Adapted with permission from Golosovsky and Solomon [72]. Copyright (2017) by the American Physical Society. https://journals.aps.org/pre/abstract/10.1103/PhysRevE.95.012324

4.1.2 Direct Citations

Figure 4.2 shows time dependence of $K_j^{dir}(t)$, an average cumulative number of direct citations for four groups of Physics papers which garnered, correspondingly, 10, 20, 30, and 100 citations in 25 years. The $K_j^{dir}(t)$ trajectories for different groups are qualitatively similar and, after scaling, all collapse onto a single curve. This is in line with Eq. 3.28.

Comparison to Eq. 3.20 yields

$$K_j^{dir}(t) = \eta_j R_0(t_j) \int_0^t \tilde{A}(\tau)d\tau, \qquad (4.1)$$

where $\tilde{A}(t)$ is given by Eq. 3.19. Collapse of the direct citation trajectories for papers with different fitness means that the aging function for citations, $\tilde{A}(t)$, does not depend on fitness. According to Eq. 3.19, this function is related to the aging function for references. Figure 4.2b validates this.

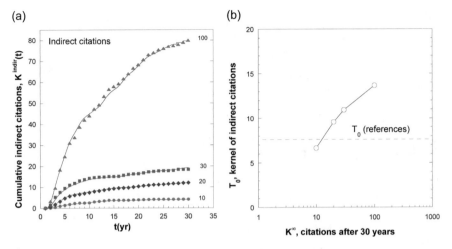

Fig. 4.3 (a) Indirect citations for 37 Physical Review B papers published in 1984. Each set of points represents cumulative indirect citations averaged over a group of papers that garnered the same number of citations K_j^∞ (10, 20, 30, and 100) by the end of 2013. Continuous lines show fits to Eq. 4.2 with $\gamma = 1.2$ yr^{-1}, $m(t - \tau)$ from Fig. 2.9, and T_{0j} as a fitting parameter for each group. (b) $T_{0j}(K_j^\infty)$ dependence. Continuous line is a guide to the eye. The dashed line shows T_0 found in our study of references. Adapted with permission from Golosovsky and Solomon [72]. Copyright (2017) by the American Physical Society. https://journals.aps.org/pre/abstract/10.1103/PhysRevE.95.012324

4.1.3   Indirect Citations

Figure 4.3 plots $K_j^{indir}(t)$, the cumulative number of indirect citations for the groups of papers shown in Fig. 4.2. We fitted these trajectories using Eq. 3.20 and the obvious relation $K_j^{indir}(t) = \int_0^t k_j(\tau)d\tau$. The fitting function was

$$k_j(t)^{indir} = \int_0^t m(t - \tau)\frac{T_j(t)}{R_0(t_j)}k_j(\tau)d\tau, \qquad (4.2)$$

where $m(t - \tau)$ is the mean number of citations for all Physics papers published in 1984 and $T_j(t) = T_{0j}e^{-\gamma_j(t-\tau)}$ is the obsolescence function which was found in our studies of references. The fitting parameters were γ_j and T_{0j}. The fitting procedure yielded the same time constant $\gamma = 1.2$ yr^{-1} for all groups of papers and equal to that found in our studies of references, as expected. However, to much of our surprise, T_{0j} turned out to be different for each group. This means that T_{0j} depends on the long-time limit of the number of citations, K_j^∞.

This unexpected $T_0(K)$ dependence indicates that our model is incomplete. The usual suspects are correlations between citation dynamics of the paper and its neighbors in citation network (in the network parlance, these are called degree-degree correlations or assortativity). Indeed, Eq. 3.18 couples citation dynamic of a paper to citation dynamics of its nearest neighbors. Our naive model of Chap. 3, assumes that these dynamics are uncorrelated, in other words, citation network lacks assortativity. However, the real citation networks exhibit assortativity [7, 72, 105]. In the next section we demonstrate that degree-degree correlations and assortativity play a significant role in citation dynamics.

4.2 Degree-Degree Correlations in the Citation Network

4.2.1 Statistics of the Second-Generation Citing Papers

Figure 4.4 schematically shows a source paper j embedded in its network neighborhood. The first-generation citing papers are those connected to it by one-hop path; the second-generation citing papers are those connected to it by two-hop paths. Any of these second-generation papers can cite the paper j indirectly.

We denote by $n_j^{nn}(t)$ and $k_j^{nn}(t)$, correspondingly, the annual numbers of the second-generation citing papers and second-generation citations per one first-generation citing paper. We also introduce $K^{nn}(t) = \int_0^t k^{nn}(\tau)d\tau$ and $N^{nn}(t) = \int_0^t n^{nn}(\tau)d\tau$, the cumulative numbers of the second-generation citations and second-generation citing papers, correspondingly. In the network parlance, K^{nn} is the average nearest-neighbor connectivity while N^{nn} is the average number of the next-nearest neighbors per one nearest neighbor. It is important to note that $K_j^{nn}(t) \geq N_j^{nn}(t)$ since each second-generation citing paper can be connected to the first-generation citing papers via several paths, as it is shown in Fig. 4.4b. While the importance of K^{nn}, the average nearest-neighbor connectivity, has been recognized

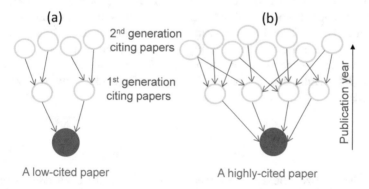

Fig. 4.4 The first- and the second-generation citing papers (nearest neighbors and next-nearest neighbors of the source paper j. (**a**) Network neighborhood of a low-cited source paper with $K = 2$ citations and the same number of citing papers. Each of these first-generation citing papers has $N^{nn} = 2$ citing papers and $K^{nn} = 2$ citations, in average. The average number of the two-hop paths connecting the second-generation citing paper to the source paper is $s = \frac{K^{nn}}{N^{nn}} = 1$. (**b**) Network neighborhood of a highly-cited source paper with $K = 4$ citations and the same number of citing papers. Each of these first-generation citing papers has $N^{nn} = 2$ citing papers and $K^{nn} = 3$ citations, in average. The average number of the two-hop paths connecting the second-generation citing paper to the source paper is $s = \frac{K^{nn}}{N^{nn}} = 1.5$

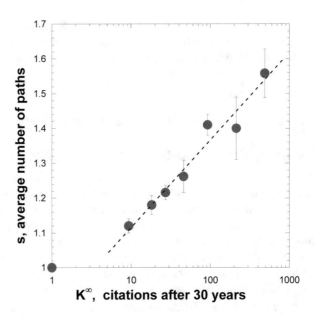

Fig. 4.5 The average number of two-hop paths connecting a second-generation citing paper to the parent paper, $s = \overline{s_j}$, where $s_j = \frac{K_j^{nn}}{N_j^{nn}}$ and the averaging is performed over each group of Physics papers published in 1984 which garnered the same number of citations by 2013. The straight line shows empirical logarithmic dependence captured by Eq. 4.4 with $\tilde{a} = 0.925$ and $\tilde{b} = 0.063$. Adapted with permission from Golosovsky and Solomon [72]. Copyright (2017) by the American Physical Society. https://journals.aps.org/pre/abstract/10.1103/PhysRevE.95.012324

in network science long ago in relation to the "friendship paradox" [7], the parameter N^{nn} has not been considered so far. Our studies show that it is equally important.

We found that K_j^{nn} and K_j are positively correlated, namely, K_j^{nn} increases with K_j, as it is schematically shown in Fig. 4.4. This means that citation network is assortative. On another hand, we found that N_j^{nn} and K_j are barely correlated, namely, N_j^{nn} hardly depends on K_j. To account for assortativity, we introduce a parameter

$$s_j = \frac{K_j^{nn}}{N_j^{nn}}, \tag{4.3}$$

which characterizes the average number of two-hop paths connecting a second-generation citing paper to its progenitor. This parameter is closely related to the so-called quadrangle coefficient [186].

Figure 4.5 shows that s slowly increases with K_j, in such a way that $s = 1$ for low-cited papers and $s = 1.55$ for highly-cited papers. This means that the former are connected to their second-generation descendants mostly by a single two-hop path, while the latter are connected to their second-generation descendants by multiple two-hop paths. The difference between the network neighborhoods of the low- and highly-cited papers may arise from the saturation effect: descendants of the low-cited papers constitute only a small fraction of all papers in the community, while descendants of highly-cited papers constitute a considerable fraction of it. We found that for $K \geq 10$, the $s(K)$ dependence shown in Fig. 4.5 may be approximated by the following expression,

$$s = \tilde{a} + \tilde{b} \ln K, \tag{4.4}$$

where $\tilde{a}$ and $\tilde{b}$ are empirical coefficients.

4.2.2 Probability of Indirect Citation

While Eq. 3.20 expresses indirect citation rate through the average annual number of the second-generation *citations*, $m(t - \tau)$, we wish to replace it by the average annual number of the second-generation *citing papers*, $n^{nn}(t - \tau)$. The rationale for this shift of the perspective is the lack of correlation between n^{nn} and K, as opposed to positive correlation between m^{nn} and K. After this replacement, Eq. 3.20 reads

$$k_j(t)^{indir} = \int_0^t n^{nn}(t - \tau) P_0 e^{-\gamma(t-\tau)} k_j(\tau) d\tau, \tag{4.5}$$

where $P_0 = s \frac{T_0}{R_0}$ is the probability amplitude of indirect citation of the source paper by a second-generation *citing paper*. This differs by factor s from $\frac{T_0}{R_0}$, the probability amplitude of indirect citation of the source paper through a second-generation *citation* which we considered in Chap. 3.

Figures 4.3b and 4.5 imply that both s and T_0 depend on K. The intuition tells us that there can't be direct relation between P_0 and the number of citations of the source paper since the citing author may be unaware of how frequently this paper has been cited. To our opinion, the relevant parameter for dynamic of indirect citations is s and not K. Indeed, after we recast the data of Fig. 4.3b as $P_0(s)$, Fig. 4.6 displays a simple linear dependence $P_0 \propto (s - s_0)$, where s_0 is the horizontal intercept.

The increasing $P_0(s)$ dependence shown in Fig. 4.6 makes much sense from the point of view of the citing author. Indeed, consider Fig. 4.7 which shows the source paper j and its first- and second-generation citing papers. The author of a can find the paper j in the reference list of the preselected paper A and cite it indirectly with probability π_1. On another hand, the author of b can find the paper j in the reference lists of *two* preselected papers, B_1 and B_2. In this case the probability of indirect citation is denoted by π_2 and it surely exceeds π_1. Not to say about paper c, which can find the paper j in the reference lists of *three* preselected papers, C_1, C_2 and C_3!

For the quantitative analysis of the probability of indirect citation in network motifs, such as those shown in Fig. 4.7, we consider a source paper j and its first- and second-generation citing papers. We denote the fraction of singlets among the second-generation citing papers as f_1, the fraction of doublets as f_2, the fraction of triplets as f_3, etc. We note that $\sum_l f_l = 1$. The average number of the two-hop paths connecting a second-generation citing paper to the source paper is $s = \sum_l l f_l$. The

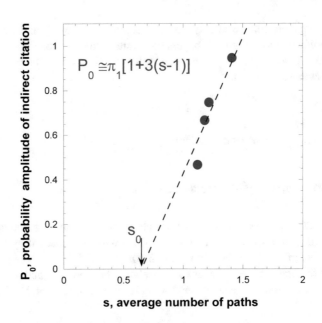

Fig. 4.6 P_0, the probability of indirect citation of the source paper by a second-generation citing paper, versus s, the average number of two-hop paths connecting these two papers. The data are for 37 papers shown in Fig. 4.3. The straight line shows a linear fit suggested by Eq. 4.9 with π_1 as a fitting parameter. Adapted with permission from Golosovsky and Solomon [72]. Copyright (2017) by the American Physical Society. https://journals.aps.org/pre/abstract/10.1103/PhysRevE. 95.012324

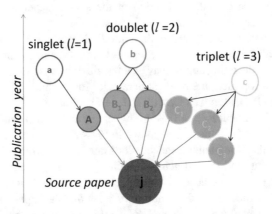

Fig. 4.7 Network motifs. The circles show the papers and the continuous lines show citations. j is the source paper, the filled circles stay for the first-generation citing papers, the open circles stay for the second-generation citing papers. Each second-generation citing paper is connected to the source paper by different number l of two-hop paths. We distinguish between network motifs such as singlet ($l = 1$), doublet ($l = 2$), triplet ($l = 3$), etc

average probability amplitude of indirect citation of the source paper by a second-generation citing paper is $P_0 = \sum_l \pi_l f_l$ where π_l is the probability amplitude for motif l.

Our measurements showed that the occurrence of motif l steeply falls as $f_l \propto 1/l^\xi$, where $\xi \approx 3$, hence in our further analysis we limit ourselves only to singlets and doublets, in other words, we assume that $f_1 + f_2 \approx 1$. Under this approximation, $s = 1 + f_2$, $P_0 = \pi_1 + f_2(\pi_2 - \pi_1)$. We exclude f_2 from these relations and come to the linear dependence

$$P_0 = (\pi_2 - \pi_1)(s - s_0) \tag{4.6}$$

with the horizontal intercept

$$s_0 = \frac{\frac{\pi_2}{\pi_1} - 2}{\frac{\pi_2}{\pi_1} - 1}. \tag{4.7}$$

Figure 4.6 shows that $P_0(s)$ dependence is indeed linear and the horizontal intercept is $s_0 = 0.6$–0.75. We use this value of s_0 and Eq. 4.7 to find the ratio of π_2 to π_1. Instead of our naive expectation, $\frac{\pi_2}{\pi_1} = 2$, Fig. 4.6 and Eq. 4.7 yield

$$\frac{\pi_2}{\pi_1} \approx 4. \tag{4.8}$$

This finding is very surprising. Indeed, consider Fig. 4.7 and the paper b which is a part of a doublet. The paper b is connected to the source paper j via two two-hop paths passing through the intermediate papers B_1 and B_2, correspondingly. Each path has a probability amplitude π_1. We would expect that the probabilities of indirect citations via these two paths sum up, namely, $\pi_2 = 2\pi_1$. This is incompatible with Fig. 4.6 and Eq. 4.8 which rather imply that $\pi_l \propto l^2$. In other words, our measurements suggest constructive interference between different citation paths. To check this conjecture, we performed microscopic measurements with three papers for which we measured f_l, the occurrence of motifs l, and the corresponding probabilities of indirect citation, π_l. Figure 4.8 shows that $\pi_l(l)$ dependence is by no means linear and is more close to quadratic, as follows from Eq. 4.8.

To include this interference effect into our model, we combine Eqs. 4.6 and 4.8 to find

$$P_0 \approx \pi_1[1 + 3(s - 1)]. \tag{4.9}$$

We substitute here $s(K)$ from Eq. 4.4 and come to

$$P_0 = a(1 + b \ln K), \tag{4.10}$$

where $a = \pi_1(3\tilde{a} - 2)$, $b = \frac{3\tilde{b}}{3\tilde{a}-2}$, and K is the number of citations of the source paper. Since K grows with time, so does P_0.

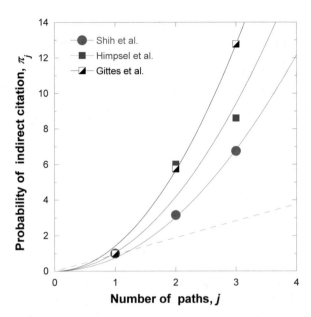

Fig. 4.8 Probability amplitude of indirect citation of the source paper by a second-generation citing paper which is a part of a network motif l, where l is the number of two-hop paths connecting the citing and cited paper. The data are for three Physical Review B papers that were published in 1984 and gained 100 citations by the end of 2013 [62, 78, 152]. The dashed line shows linear dependence, $\pi_l \propto l$, which would occur in the absence of interference, namely, if the probabilities of different paths just sum up. All data for $l \geq 2$ lie well-above the dashed line corresponding to this linear dependence. Continuous lines show quadratic dependences, $\pi_l \propto l^2$. All this suggests constructive interference between different citation paths. Adapted with permission from Golosovsky and Solomon [72]. Copyright (2017) by the American Physical Society. https://journals.aps.org/pre/abstract/10.1103/PhysRevE.95.012324

Chapter 5
Model Validation

Abstract In Chap. 4 we presented the calibration procedure, namely, the measurements of the empirical functions and parameters of the model. These measurements were focused on the deterministic component of citation dynamics while the stochastic component was averaged out. Here, we focus on the fluctuating component of citation dynamics of individual papers and verify that it is captured by the model as well.

Keywords Citation distribution · Fitness · Autocorrelation

5.1 Numerical Simulation of Stochastic Model

For readers convenience, we summarize here our model. The latent citation rate of a paper j in year t after publication is

$$\lambda_j(t) = \eta_j R_0(t_j)\tilde{A}(t) + \int_0^t n^{nn}(t-\tau)P_0 e^{-\gamma(t-\tau)}k_j(\tau)d\tau, \tag{5.1}$$

where η_j is the paper's fitness, an empirical parameter, unique for each paper; $\tilde{A}(t)$ is the aging function for citations; P_0 is the probability amplitude of indirect citation of the paper j by a second-generation citing paper; γ is the obsolescence exponent; and $n^{nn}(t)$ is the mean annual number of the second-generation citing papers per one first-generation citing paper. Citation rate of the paper is $k_j(t) = \Delta K_j/\Delta t$, where ΔK_j is the number of citations garnered in year t and $\Delta t = 1$ year. The model assumes that ΔK_j follows the Poisson distribution, $\Delta K_j = \frac{\lambda_j^{\Delta K_j}}{\Delta K_j!}e^{-\lambda_j}$. The exponent γ and the functions $\tilde{A}(t)$ and $n^{nn}(t)$ are the same for all papers in one field published in the same year, the function $P_0(K)$ is given by Eq. 4.10. Our specific goal here is to validate not only the average component of citation dynamics, as it was done in Chap. 4, but its stochastic component as well.

5.1.1   Methodology

If citation dynamics of individual papers were following a homogeneous stochastic process, the comparison to the model could be performed on the paper-by-paper basis. Namely, we would decompose each citation trajectory onto the mean and fluctuating parts and compare them to the model, as it is schematically shown in Fig. 5.1.

However, citation dynamic of a paper is a non-homogeneous discrete stochastic process which can't be easily decomposed into deterministic and stochastic components. To perform a meaningful comparison to the model, we adopted the following strategy. We considered a large set of papers in one field, that were published in the same year, and measured their citation dynamics using the Web of Science database. In the framework of our model, the papers in this set differ only by their fitness. Then, we designed a synthetic set with the same number of papers and the same fitness distribution and performed numerical simulations of their citation dynamics basing on the model. Comparison between the actual and synthetic sets was performed along several dimensions:

1. Cumulative citation distributions.
2. Citation trajectories of individual papers.
3. Autocorrelation of citation trajectories.
4. The mean and the fluctuating components of citation dynamics.
5. The number of uncited papers.

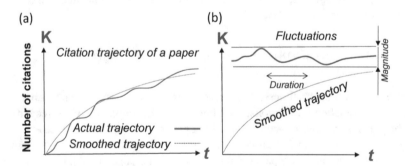

Fig. 5.1 (**a**) Schematic representation of the citation trajectory of a paper. (**b**) Decomposition onto the smooth part (mean trajectory) and fluctuations. The fluctuations are characterized by their magnitude and duration. This idealized scheme would work if citations could be represented by a continuous variable. Since citations are discrete, this simple scheme should be modified

5.1.2 Citation Distributions

Figure 5.2 shows citation distributions for the set of 40,195 Physics papers published in 1984 [72]. To simulate these distributions using our model, one shall know the fitness of each paper. In principle, it can be measured by tracing dynamics of direct citations. Indeed, once we know the cumulative number of the direct citations, $K_j^{dir}(t)$; the fitness of the paper j can be found from the expression $\eta_j \approx \frac{K_j^{dir}(t)}{\int_0^t \tilde{A}(\tau)d\tau}$. However, to measure $K_j^{dir}(t)$ for every paper proved to be too complicated, hence we found the fitness distribution in an alternative way. The basis of our strategy is the observation that there is a delay between the publication of a paper and the onset of indirect citations (Fig. 4.1), hence citations garnered by a paper during first 2–3 years after publication are mostly direct. Therefore, for the set of papers published in the same year, citation distributions 2–3 years after publication mimic the fitness distribution (within a constant factor). We found, that these early citation distributions are close to log-normal. Having this in mind, we run the model simulation (Eq. 5.1) for 40,195 papers with a log-normal fitness distribution, $\rho(\eta) = \frac{1}{\sqrt{2\pi}\sigma\eta}e^{-\frac{(\ln\eta-\mu)^2}{2\sigma^2}}$, where μ and σ were the fitting parameters. Our aim was to achieve the best fit to the measured citation distribution for $t = 2$.

Other parameters of the simulation were as follows. We used $\gamma = 1.2\ \text{yr}^{-1}$, as found in our measurements of indirect references and citations; $\tilde{A}(t)$ from

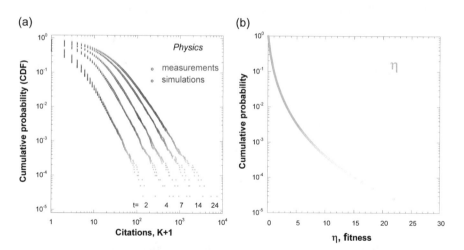

Fig. 5.2 (a) Annual cumulative citation distributions for 40,195 Physics papers published in 1984. The red circles show measured data, the blue circles show results of stochastic simulation based on the Poisson process with the rate given by Eq. 5.1. (b) Fitness distribution used for simulation. It is a log-normal distribution with $\mu = -1.48$ and $\sigma = 1.12$. The fitness distribution is so wide that it includes the papers with the fitness which exceeds the average one by the factor of 70. Adapted with permission from Golosovsky and Solomon [72]. Copyright (2017) by the American Physical Society. https://journals.aps.org/pre/abstract/10.1103/PhysRevE.95.012324

Fig. 4.2b, and $P_0(K)$ from Eq. 4.10. We assumed that $n^{nn}(t)$ and $m(t)$ dependences (see Fig. 2.9) are identical, namely $n^{nn}(t) = \frac{m(t)}{\bar{s}}$ where $\bar{s} = 1.2$ is the average number of two-hop paths connecting the second-generation citing paper to its progenitor. Figure 5.2 demonstrates an excellent agreement between the measured and simulated citation distributions.

5.1.3   Citation Trajectories

Although the model captures citation distributions perfectly well, such validation is still insufficient since different sets of citation trajectories can yield the same set of citation distributions. To further validate our model, we compared the measured and simulated citation trajectories. It should be noted, that citation dynamics of papers follows a self-exciting Hawkes process which amplifies past fluctuations. Therefore, even for the same initial conditions, the spread of simulated citation trajectories is so wide that the comparison of the measured and simulated trajectories on the paper-by-paper basis is meaningless. Therefore, we compared citation trajectories for the sets of similar papers.

Figures 5.3 and 5.4 show citation trajectories for the sets of papers that garnered the same number of citations in the long-time limit. If we perform averaging for each set, the measured and simulated average citation trajectories agree well. This is not unexpected since the empirical function for the probability of indirect citations, $P_0(K)e^{-\gamma(t-\tau)}$, was established just from the requirement that the model fits the

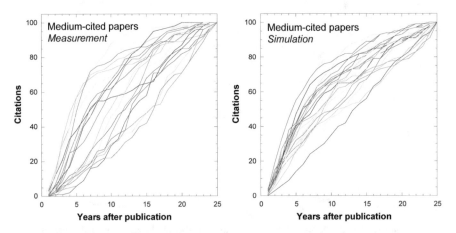

Fig. 5.3 Citation trajectories of the Physics papers that were published in 1984 and accrued 99 citations in subsequent 25 years. Stochastic numerical simulation based on our model correctly predicts the shape and the spread of citation trajectories. Adapted with permission from Golosovsky and Solomon [72]. Copyright (2017) by the American Physical Society. https://journals.aps.org/pre/abstract/10.1103/PhysRevE.95.012324

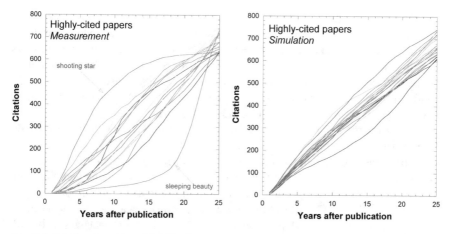

Fig. 5.4 Citation dynamics of the Physics papers that were published in 1984 and accrued 600–750 citations in subsequent 25 years. For most papers, the model correctly predicts the shape and the smoothness of citation trajectories. However, the model does not capture well extreme cases such as a sleeping beauty—the paper with delayed recognition, or a shooting star—the paper that is highly popular at the beginning of its citation career but interest to it fades away soon. Adapted with permission from Golosovsky and Solomon [67]. Copyright (2017) by the American Physical Society. https://journals.aps.org/pre/abstract/10.1103/PhysRevE.95.012324

average citation dynamics of similar papers. Different sets of trajectories can have the same average trajectory, hence comparison of the shapes of the measured and simulated citation trajectories of individual papers tells a separate story, beyond the average.

Figure 5.3 shows that for low- and moderately-cited papers, the measured and simulated trajectories look very similar—they exhibit fluctuations which have more or less the same size and duration. Figure 5.4 shows that for highly-cited papers, both sets of trajectories are smooth, but the spread of the measured trajectories exceeds that of the simulated ones. Although simulation yields a substantial spread in citation trajectories, it does not reproduce well such extreme cases as "sleeping beauty" or "shooting star". This is rewarding, because the unpredictability of science is one of the most important factors that makes it fun. (In fact, the model reproduces "sleeping beauties" but they are very rare, while in reality they are much more frequent.)

5.2   Stochastic Component of Citation Dynamics

5.2.1   Methodology

Although the agreement between the measured and simulated citation distributions is impressive, this fact alone is not sufficient to validate the model. The same distributions could result from very different citation trajectories. For example, the jerky and the smooth citation trajectories can yield the same set of citation distributions. So, we performed an additional test and verified that the model correctly reproduces not only citation distributions at different times but it reproduces citation trajectories of individual papers as well. Figures 5.3 and 5.4 illustrate that, for the sets of similar papers, the model correctly reproduces the spreads of the measured and simulated citation trajectories. In what follows, we analyze how well the model reproduces fluctuations, namely, the jerkiness of citation trajectories.

To measure the fluctuating component of the citation dynamics of papers, we do not analyze citation trajectories of individual papers with the aim of decomposing them onto the smooth and fluctuating components, since citations are discrete and these two components of the citation trajectory are inseparable (see Fig. 5.1). Moreover, since citation dynamic is a time-inhomogeneous stochastic process, the smoothing of the citation trajectory of a paper is not well-defined. To meet these difficulties, we adopted a different approach and aimed at statistical characterization of citation dynamics of the sets of similar papers. Indeed, in the framework of our model, the key individual parameter that determines citation dynamics of a paper is its fitness. For a set of research papers in one discipline, published in the same year, all having the same fitness, we analyzed statistics of citations at a fixed time. Using a series of snapshots of the citation trajectories of this set we measured statistical properties of additional citations and found their mean, variance, and autocorrelation. The variance is a measure of the magnitude of fluctuations while the autocorrelation coefficient gauges their duration.

The difficulty here is how to find the papers with the same fitness. To this end, we had to analyze citation trajectory of every paper and to compare it to the model. We took a shortcut and assumed that the papers in one field which were published in the same year, and garnered the same number of citations by a certain time t, have the same fitness.

5.2.2   Statistical Distribution of Additional Citations

5.2.2.1   Mean

To characterize the smooth part of citation dynamics, we used the same set of papers as in Fig. 5.2, namely, we sorted Physics papers published in 1984 into the bins containing the papers with the same $K(t)$, the number of citations garnered during t years after publication. For the papers in each bin, we considered statistical distribution of additional citations, $k_j = \Delta K_j$ (citations garnered in the year $t + 1$). This was done for the measured and simulated data as well. Figure 5.5 shows that the mean number of additional citations $\overline{k_j}$ for the measured and simulated distributions are very close. From this we conclude that the model captures fairly well the smooth part of citation dynamics of papers.

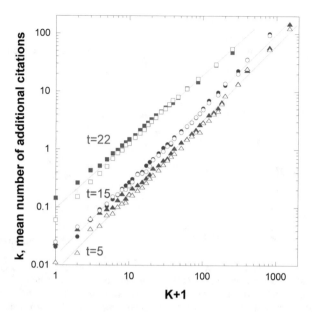

Fig. 5.5 The mean number of additional citations, $\overline{k}(t)$, in dependence of the number of previous citations, $K(t)$. The averaging is performed over all papers which were published in the same year and garnered the same number of citations K during t years after publication. The filled symbols show measured $\overline{k}(t)$, the open symbols show results of the numerical simulation. Note good correspondence between the measurement and simulation. The straight lines show the fit to the preferential attachment model, $\overline{k}(t) \propto (K + K_0)^\zeta$ with $\zeta = 1.1$–1.25 and $K_0 = 1$ (see Chap. 9). The data are for 40,195 Physics papers published in 1984

5.2.2.2   Autocorrelation

To characterize the duration of the citation fluctuations, we considered the autocorrelation between additional citations acquired by a paper in subsequent years. We calculated it for the annual citations measured for the set of papers which have the same number of previous citations $K(t)$. Specifically, we determined the number of citations garnered by each paper during two subsequent years, $k_j(t)$ and $k_j(t-1)$, and calculated the Pearson autocorrelation coefficient for this set,

$$c_{t,t-1} = \frac{\overline{\left(k_j(t) - \overline{k}_j(t)\right)\left(k_j(t-1) - \overline{k}_j(t-1)\right)}}{\sigma_t \sigma_{t-1}}, \tag{5.2}$$

where $\overline{k}_j(t)$ and $\overline{k}_j(t-1)$ are, correspondingly, the mean of the $k_j(t)$ and $k_j(t-1)$ distributions, and $\sigma(t)$ and $\sigma(t-1)$ are standard deviations.

   What is the physical meaning of $c_{t,t-1}$? If citations were following the Markov process, then citation rate of a paper should be determined by the total number of accumulated citations and independent of the previous citation history, in such a way that $c_{t,t-1} = 0$. On another hand, $c_{t,t-1} = 1$ indicates that citation rate of a paper is

(a)   (b)

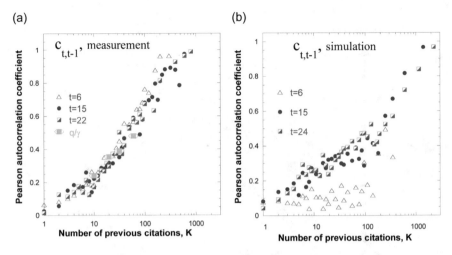

Fig. 5.6 The Pearson autocorrelation coefficient for additional citations $k_j(t)$ and $k_j(t-1)$. Each point corresponds to the set of papers with the same number of previous citations K garnered after t years. (**a**) Measurements. (**b**) Numerical simulation. The simulation agrees with the measurements for $t > 10$ and underestimates $c_{t,t-1}$ for $t < 10$. Green symbols in (**a**) show prediction of Eq. 5.4 based on Fig. 4.3b. We observe that $c_{t,t-1}(K)$ dependence mimics the $q(K)$ dependence

fully determined by the number of citations garnered last year. Thus, $c_{t,t-1}$ gauges the deviation of the citation process from the Markovian. From another perspective, $c_{t,t-1} = 0$ indicates that citation rates in subsequent years are uncorrelated, namely, characteristic duration of the fluctuations is less than 1 year. On another hand, $c_{t,t-1} = 1$ indicates strong positive correlation between citation rates of a paper in subsequent years.

Figure 5.6 shows that $c_{t,t-1}$ is not the same for all papers but grows with the number of accumulated citations K. For low-cited papers, $c_{t,t-1}$ is small. This indicates that citation dynamics of these papers can be described by a Markov process. This is in contrast to the highly-cited papers, for which $c_{t,t-1} \sim 1$, indicating that citation dynamics of these papers are heavily influenced by their citation history and do not follow a Markov process. These considerations also explain the jerky citation trajectories of the low-cited papers as opposed to smooth trajectories of the highly-cited papers (Figs. 5.3a and 5.4a).

Our model reproduces measured $c_{t,t-1}(K)$ dependencies fairly well for $t > 10$, although for $t < 10$ the simulated $c_{t,t-1}(K)$ values are smaller than the measured ones.

Dramatic difference between citation dynamics of the low- and highly-cited papers (Markovian versus non-Markovian process) is a direct consequence of nonlinearity in Eq. 5.1 which stems from the $P_0(K)$ dependence. We claim that the $c_{t,t-1}(K)$ dependence shown in Fig. 5.6 stems from the same source. To show this, we go to continuous approximation of the model (Eq. 3.25) and notice that since $1/\tilde{\gamma} \approx 0.8$ is close to 1 year, then this equation can be recast as

$$k_j(t) \approx \eta_j R_0 \tilde{A}(t) + \frac{q_j}{\tilde{\gamma}} k_j(t-1). \tag{5.3}$$

This equation contains $k_j(t)$ and $k_j(t-1)$ which are closely related to $c_{t,t-1}$. To exclude η_j from Eq. 5.3, we recall Eq. 3.28 and write $\eta_j = \frac{K_j(t)}{B(t)} = \frac{K_j(t-1)}{B(t-1)}$. We note that $K_j(t-1) = K_j(t) - k_j(t-1)$ where $K_j(t) = K$ for all papers in the set. We substitute these expressions into Eqs. 5.3, 5.2 and find $c_{t,t-1} \approx \left(\frac{q}{\gamma} - \frac{\tilde{A}(t)}{B(t-1)} \right) \frac{\sigma_{t-1}}{\sigma_t}$. For $t > 10$, the second term in the parentheses is negligible. In addition, $\frac{\sigma_{t-1}}{\sigma_t} \approx 1$. Finally, we find that

$$c_{t,t-1} \approx \frac{q}{\gamma} \tag{5.4}$$

and does not depend on t. Indeed, Fig. 5.6a shows that $c_{t,t-1}$ dependences for different years collapse onto a single curve. Thus, the autocorrelation coefficient is a direct measure of the parameter q which is directly related to the probability of indirect citations, P_0. Indeed, Fig. 5.6 shows that the $c_{t,t-1}(K)$ dependence mimics the $q(K)$ dependence.

5.2.2.3   Variance

To measure the fluctuating part of the citation dynamics of papers, we considered the variance of the statistical distribution of additional citations k_j for similar papers, namely, those that garnered the same number of citations $K(t)$ during the same time period t. This variance comes from two sources: (1) intrinsic variance resulting from the probabilistic character of citation process, and (2) variability of citation dynamics of the papers in this set which is associated with their different citation history. The latter source of variability arises from the fact that citation dynamic of a paper is not a Markov process, it depends not only on the accumulated number of citations K, but on the whole citation history of a paper.

Figure 5.7 schematically illustrates these two sources of variability. We consider a set of papers which were published in the same year and which garnered the same number of citations by year t. The width of the statistical distribution of additional citations $k_j(t)$ is determined by the sum of intrinsic randomness (homogeneous broadening) and that resulting from the spread of the recent citation history of these papers (inhomogeneous broadening). To exclude the the latter, we performed autoregression, namely we subtracted the history-dependent part of additional citations [70]. In particular, for each paper j we considered the autoregressive variable

$$X_j(t) = k_j(t) - c_{t,t-1} \frac{\sigma_t}{\sigma_{t-1}} \left[k_j(t-1) - \bar{k}_j(t-1) \right], \tag{5.5}$$

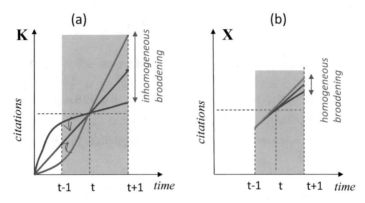

Fig. 5.7 (**a**) Schematic representation of citation trajectories of papers that were published in the same year and garnered the same number of citations after t years. The width of the statistical distribution of their citations garnered in the year $t + 1$ (additional citations) is determined by the intrinsic randomness of the citation process (homogeneous broadening) and by the recent citation history of these papers (inhomogeneous broadening). (**b**) The autoregression procedure (Eq. 5.5) subtracts the history-related contribution and leaves only intrinsically random component of the additional citations

where $c_{t,t-1}$ is the Pearson autocorrelation coefficient and $k_j(t-1)$, $k_j(t)$ are the numbers of additional citations garnered, correspondingly, in years $t-1$ and t. Obviously, $\overline{X_j} = \overline{k_j}$. However, the variance of X_j should be smaller than the variance of k_j, since the history-related contribution has been largely subtracted. We measured the variance of X_j and considered the variance-to-mean ratio, the so-called Fano number, $F = \frac{(X_j - \overline{X_j})^2}{\overline{X_j}}$. If $k_j(t)$ dependence on citation history were fully captured by the first-order autoregressive Eq. 5.5, then $X_j(t)$ shall be history-independent and its statistics shall be close to Poissonian, for which $F = 1$.

Figure 5.8 plots the variance of X_j versus mean. This is done both for measured and simulated data. Let's analyze first the results of the simulation (open circles). The data for low- and medium-sized papers lie on the $F = 1$ line corresponding to the Poissonian distribution. The upward deviation for $k > 10$ (corresponding to highly-cited papers that garnered more than $\sim$200 citations after 15 years) shows the limits of the approximation of the history-dependent part of citation dynamics through the first-order autoregressive Eq. 5.5.

As for the measured data, Fig. 5.7 shows that for moderately- and low-cited papers with $\overline{k_j} < 1$, our measurements closely follow the numerical simulation. This demonstrates a good agreement between the measurements and the model and indicates that the stochastic component of the citation dynamics of the low-cited papers is indeed Poissonian. In terms of our model, this proves that the parameters of the model for all these papers are identical and the fluctuations come not from the spread of the parameters for different papers but from the shot noise associated with discrete nature of citations. From another perspective, since citation dynamics of the

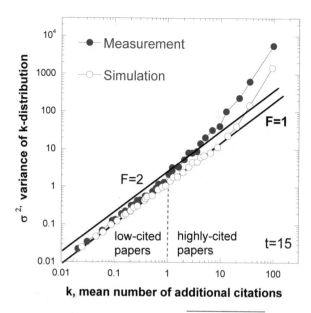

Fig. 5.8 Variance of the additional citations, $\sigma^2(t) = \overline{\left(X_j(t) - \overline{X_j}(t)\right)^2}$, versus mean, $\overline{X_j}(t) = \overline{k_j}(t)$. The filled circles show measured values, the empty circles show results of numerical simulation, the straight lines indicate constant variance-to-mean ratio (Fano number) where $F = 1$ corresponds to the Poisson distribution. Note good correspondence between the measured and simulated data for low- and moderately-cited papers and deviation for highly-cited papers. The data are for 40,195 Physics papers published in 1984

low-cited papers is determined mostly by direct citations, this further validates our assumption that the aging function $A(t)$ is the same for all papers.

On another hand, the measured variance for the highly-cited papers exceeds that obtained in simulations. This indicates on the limitation of the model which assumes the same parameters of citation dynamics for the papers with the same $K(t)$. For these papers, indirect citations play an increasingly important role and the upward deviation from the simulation indicates on the spread of the model parameters for indirect citations. We believe that this variability comes mostly from the parameter k^{nn} which characterizes the nearest-neighbor connectivity in the citation network and can vary in broad limits [56].

5.2.3   Uncited Papers

Figure 5.9 shows that our model correctly predicts the fraction of uncited papers at every year after publication. The good correspondence between the measured number of uncited papers and the model prediction indicates that uncited papers are a natural outcome of citation process [27].

Fig. 5.9 Time dependence of
the fraction of the Physics
papers that remained uncited
25 years after publication.
The data is for 40,195
Physics papers published in
1984. Note good agreement
between the measurement and
simulation

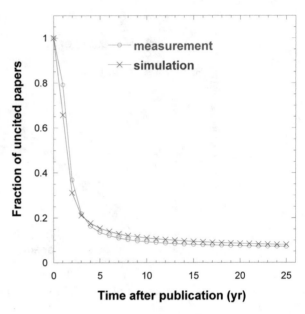

Chapter 6
Comparison of Citation Dynamics for Different Disciplines

Abstract We demonstrate our measurements of citation dynamics of the Physics, Mathematics, and Economics papers. Our model captures them very well. We discuss the similarity and distinctions between citation dynamics of the papers belonging to different disciplines.

Keywords Citation dynamics · Citation distribution · Citation lifetime · Longevity · Uncitedness · Universality

6.1 Extension of the Model to Other Disciplines

How general is the model of citation dynamics presented in Chap. 3? While it was calibrated using Physics papers, we show here that it can be extended to other disciplines as well, albeit with different parameters. For readers convenience, we recapitulate here the model. It assumes that citation dynamic of a paper j published in year t_j follows an inhomogeneous Poisson process with the rate

$$\lambda_j(t) = \eta_j R_0 \tilde{A}(t) + \int_0^t n^{nn}(t-\tau) P_0 e^{-\gamma(t-\tau)} k_j(\tau) d\tau. \tag{6.1}$$

The first addend in this equation captures the direct citations. η_j is the papers's fitness which is drawn from the fitness distribution $\rho(\eta)$, R_0 is the average length of the reference list of the papers published in year t_j, t is the number of years after publication, and $\tilde{A}(t)$ is the aging function for citations which is related to the aging function for references $A(t)$ through relation $\tilde{A}(t) = A(t)e^{(\alpha+\beta)t}$ where $\int_0^\infty A(\tau)d\tau = 1$ and the exponents α and β characterize the growth of the number of publications and of the reference list length, correspondingly.

The second addend in Eq. 6.1 captures the indirect citations. $k_j(\tau)$ is the number of citations garnered by the paper in year $t_j + \tau$, $n^{nn}(t)$ is the mean annual number of the second-generation citing papers per one first-generation citing paper, γ

© The Author(s), under exclusive license to Springer Nature Switzerland AG 2019
M. Golosovsky, *Citation Analysis and Dynamics of Citation Networks*,
SpringerBriefs in Complexity, https://doi.org/10.1007/978-3-030-28169-4_6

is the exponent of the obsolescence function, P_0 is the probability amplitude of indirect citation which depends on the number of accumulated citations, $K_j(t) = \int_0^t k_j(\tau)d\tau$, as follows

$$P_0(K) = a(1 + b \ln K), \tag{6.2}$$

where a, b, and γ are empirical parameters.

Equations 6.1 and 6.2 state that citation dynamic of a paper is determined, on the one hand, by the paper-specific parameters and variables: fitness η_j, number of annual $k_j(t)$ and accumulated citations $K_j(t)$; on another hand, it is determined by the empirical parameters and functions R_0, α, β, γ, a, b, $A(t)$, and $n^{nn}(t)$ which are common for all papers published in the same year and belong to one discipline. To find these parameters, we performed measurements of citation dynamics of all Physics, Mathematics, and Economics papers published in the same year. The results for Physics papers have been demonstrated in Chap. 5, here we focus on our measurements with Economics and Mathematics papers. We show that citation dynamics of the papers belonging to different disciplines are qualitatively similar and differ mainly in scale which is set by R_0—the average reference list length. This scale is specific for each discipline and citation dynamics of papers depend on it in a non-trivial way.

6.1.1 Mean Number of Citations

We start our consideration of citation dynamics from the mean annual number of citations, $m(t)$, and the corresponding cumulative mean, $M(t) = \int_0^t m(\tau)d\tau$. Figure 6.1 shows that $M(t)$ dependences for three disciplines look very different: some of them show signs of saturation while others do not. We claim that divergence of the $M(t)$ dependences in the long time limit is related to the exponential growth of the number of publications and of the reference list length. To compensate for these factors, we considered

$$M_{detrended}(t) = R(t) = \int_0^t m(\tau)e^{-(\alpha+\beta)\tau}d\tau, \tag{6.3}$$

where α and β are the growth exponents for the number of publications and for the reference list length, correspondingly (see Chap. 2, Sect. 2.3), and $R(t)$ is the age distribution of references. In contrast to $M(t)$, which can diverge in the long time limit, $R(t) = R_0 \int_0^t r(\tau)d\tau$ and converges to R_0 in the long time limit. Here, $r(t)$ is the reduced age distribution of references.

Basing on Eq. 6.3, by trial and error, we found the growth exponents for each discipline from the condition that $R(t)$ achieves saturation in the long-time limit. For Physics, Economics, and Mathematics, this procedure yielded, correspondingly, $\alpha + \beta = 0.04, 0.085, 0.092$ yr^{-1}; and $R_0 = 22, 9.1, 4.7$. The R_0 value for Physics corresponds to our independent measurements, while the R_0 values for Economics and Mathematics are, probably, too small. The reason for this is that these numbers

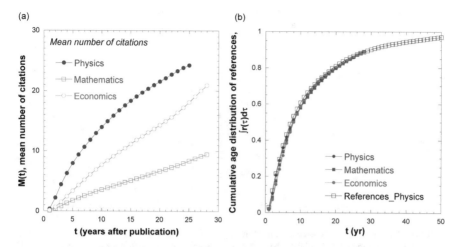

Fig. 6.1 The mean number of citations for 40,195 Physics, 6313 Mathematics, and 3043 Economics papers, all published in 1984. (**a**) Raw data. For Physics papers, $M(t)$ grows with deceleration and probably comes to saturation, while for Mathematics and Economics papers, $M(t)$ exhibits a permanent growth with no signs of saturation. (**b**) Detrended and scaled data, $\int_0^t r(\tau)d\tau$, where $r(t) = \frac{R(t)}{R_0}$, according to Eq. 6.3. All $r(t)$ curves are nearly identical. Adapted with permission from Golosovsky [67]. Copyright (2017) by the American Physical Society. https://journals.aps.org/pre/abstract/10.1103/PhysRevE.96.032306

do not include interdisciplinary references and books. The latter constitute a very small part of Physics references while they are abundant among Mathematics and Economics references.

Figure 6.1 shows the $\int_0^t r(\tau)d\tau$ dependences for each discipline found from Eq. 6.3. Our model does not require them to collapse onto one universal dependence but, in fact, they nearly do this.

6.1.2 Citation Distributions and the Aging Function for Citations

We consider now citation distributions for different disciplines (Figs. 5.2 and 6.2). According to Eq. 6.1, citation distributions during first 2–3 years after publication are determined by the first addend (direct citations) while the second addend (indirect citations) takes the lead later on. Having this in mind, we analyzed citation distributions early after publication and determined the corresponding fitness distributions. For all three disciplines, we found log-normal fitness distribution, $\rho(\eta) = \frac{1}{\sqrt{2\pi}\sigma\eta}e^{-\frac{(\ln\eta-\mu)^2}{2\sigma^2}}$, with nearly the same parameters: $\mu \approx -1.48$, $\sigma = 1.05$–1.12.

With respect to the aging function for citations $\tilde{A}(t)$, we found it by fitting the whole citation distribution at each t using our model. Figure 6.3 shows that

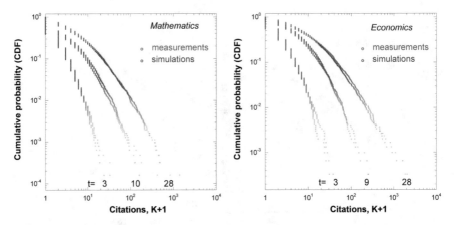

Fig. 6.2 Annual cumulative citation distributions. Red circles show measured data, blue circles show results of stochastic simulation based on the Poisson process with the rate given by Eq. 6.1. Left panel: 6313 Mathematics papers published in 1984. Right panel: 3043 Economics papers published in 1984. t is the time after publication where publication year corresponds to $t = 1$. The corresponding distributions for Physics papers are shown in Fig. 5.2. Adapted with permission from Golosovsky [67]. Copyright (2017) by the American Physical Society. https://journals.aps. org/pre/abstract/10.1103/PhysRevE.96.032306

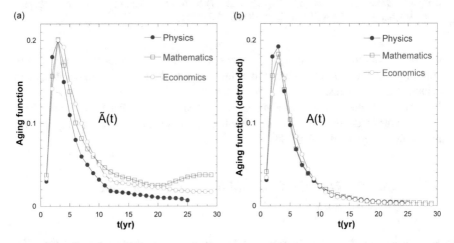

Fig. 6.3 (a) $\tilde{A}(t)$, aging functions for citations as found from the analysis of citation distributions. $\tilde{A}(t)$ for different disciplines are not the same. (b) $A(t)$, aging function for references determined from the relation $A(t) = \tilde{A}(t)e^{-(\alpha+\beta)t}$ where the exponents were found from Fig. 6.1. All three $A(t)$ dependences collapse onto a single curve suggesting that the strategy, which the authors use to find references for their papers, is nearly universal

the aging functions vary from one discipline to another. This difference can be traced to the discipline-specific growth exponents, α and β. Indeed, the detrended function $\tilde{A}(t)e^{-(\alpha+\beta)t}$ is nothing else but the aging function for references $A(t)$. We substituted in this expression $\alpha + \beta$ determined from Fig. 6.1 and found $A(t)$

for each discipline. Surprisingly, all three $A(t)$ dependences collapse onto one curve. Although in the framework of our model, $A(t)$ dependence could be specific for each discipline, Fig. 5.9 shows that, in fact, it is universal. This goes hand in hand with the observation that the reduced age distribution of references is nearly universal (Fig. 6.1b). All this suggests that the strategy, which the authors use to find references, is more or less universal.

6.1.3 Indirect Citations

To find the kernel for indirect citations, we sought to reproduce not only citation distributions but the autocorrelation coefficient $c_{t,t-1}$ as well (see Chap. 5, Eq. 5.2). To this end, we used Eq. 6.1 with already found empirical functions $\rho(\eta)$ and $\tilde{A}(t)$, the fitting parameters being a, b, and γ. The simulation fairly well reproduces the measured citation distributions at all times (Fig. 6.2) and, for all three disciplines, yields almost the same obsolescence exponent, $\gamma \approx 1.2$ yr^{-1}. However, the parameters a and b turned out to be discipline-dependent. Figure 6.4 shows that these parameters depend systematically on R_0, the average reference list length. To our opinion, this is expected since R_0 is a scale for the number of citations for each discipline.

Consider first the $a(R_0)$ dependence. We claim that it stems from Eq. 3.20 which states that the probability of choosing the indirect citation from a preselected paper inversely depends on the length of its reference list. If we compare disciplines with different R_0, we come to conclusion that $P_0 \propto 1/R_0$. This implies that $a \propto 1/R_0$ and Fig. 6.4 proves this conjecture. Thus, in the issues of the reference list composition, the behavior of the authors in different disciplines is very similar.

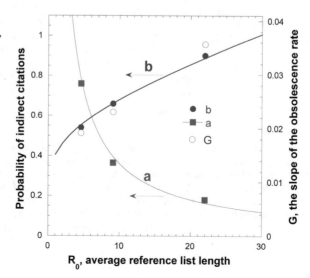

Fig. 6.4 Probability amplitude of indirect citation, $P_0(K) = a(1 + b \ln K)$. The filled symbols show the parameters a and b found by fitting citation distributions (Fig. 6.2) using Eqs. 6.1 and 6.2. The red continuous line is a $a \propto \frac{1}{R_0}$ dependence suggested by Eq. 3.20. The blue continuous line shows prediction of Eq. 6.4 where $K_{crit} = \frac{Q}{R_0}$ and Q is the community size. The open circles show G, the slope of the obsolescence rate found from Fig. 6.6b and Eq. 6.5

What could be a reason for the $b(R_0)$ dependence? In Chap. 4 we found that the parameter b is related to the structure of citation network: it characterizes the average number of two-hop paths connecting a second-generation citing paper to its progenitor. To understand these issues more deeply, we consider a paper and its contemporary community, namely, all papers that were published afterwards and which can, in principle, cite it (Fig. 6.5). The second-generation descendants of a low-cited paper constitute only a small part of this community, hence they are connected to their progenitor by a single two-hop path. This is in contrast to the second-generation descendants of a highly-cited paper which encompass the whole community. Many of them are connected to its progenitor by multiple two-hop paths.

To put these considerations onto a quantitative basis, we consider some source paper and all its second-generation citing papers. The number of citations of this paper is K, while M^{nn}, N^{nn} are, correspondingly, the numbers of the second-generation citations and citing papers per one first-generation citing paper. The total number of the second-generation citations and citing papers is $K M^{nn}$ and $K N^{nn}$, correspondingly. The ratio of these numbers yields s, an average number of two-hop paths connecting the source paper to a second-generation citing paper. Figure 6.5a shows that for a low-cited paper, whose descendants comprise a small part of its community, $N^{nn} = M^{nn}$ and $s = 1$. On another hand, for a highly-cited paper, whose descendants encompass the whole community, Fig. 6.5b shows that $K N^{nn} = Q$ and $s = \frac{K M^{nn}}{Q} > 1$. The critical number of citations that delineates between the low- and highly-cited papers in this context is $K_{crit} = \frac{Q}{M^{nn}}$. For a moment, we disregard network assortativity and assume that $M^{nn} = M$. According

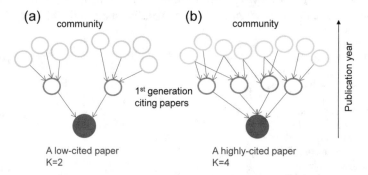

Fig. 6.5 A paper and its community, namely, all papers that can potentially cite it. K is the number of citations, $K N^{nn}$ is the number of its next-nearest neighbors, $M = M^{nn}$ is the average number of citations, and Q is the size of community. In this example $M = 3$ and $Q = 8$ for both panels. (**a**) A low-cited paper. $K = 2$, $K N^{nn} = 6$, in such a way that $K M < Q$, and each next-nearest neighbor is connected to the source paper by a single path. (**b**) A highly-cited paper. $K = 4$, $K N^{nn} = Q = 8$ while the number of the second-generation citations is $K M = 12$. Therefore, the average number of paths connecting the source paper to its next-nearest neighbors is $\frac{K M}{Q} = 1.5$

to the duality principle, M scales with R_0, the average length of the reference list, hence $K_{crit} \propto \frac{Q}{R_0}$.

How K_{crit} enters our model? Since Eq. 6.2 depends not on K but on $\ln K$, we conjecture that the general expression for the probability amplitude of indirect citation is

$$P_0 = a \left(1 + \frac{\ln K}{\ln K_{crit}} \right), \tag{6.4}$$

in other words, $b = (\ln K_{crit})^{-1}$. The dashed line in Fig. 6.4 shows that Eq. 6.4 captures our data fairly well and yields the same $Q \approx 200$ for all three disciplines. According to our interpretation, this is the number of active working groups in the community. This is remindful of Dunbar's number, namely, the maximum size of a coherent human group.

6.2 Citation Lifetime

We found that citation dynamics of scientific papers are nonlinear, the nonlinearity is captured by the logarithmic term in Eq. 6.4. Although the nonlinearity is weak, it has important consequences, the most spectacular one being the presence of runaways, namely, the papers whose citation career does not come to saturation and which continue to be cited indefinitely. To analyze this phenomenon, we studied citation lifetime of papers. To this end, we approximated citation trajectory of each paper by the exponential dependence, $K = K^\infty \left(1 - e^{-\frac{t-\Delta}{\tau_0}} \right)$, where τ_0 is the citation lifetime, K^∞ is the number of citations in the long time limit, and $\Delta \sim 1$–2 years characterizes delay in the onset of the citation career of a paper. Figure 6.6a shows that τ_0 grows continuously with K and diverges at certain K_r, in such a way that the papers with $K > K_r$ become supercritical [14], in other words, they exhibit runaway behavior—their citation career does not saturate. Similar runaways were detected in the distribution of the Web page popularity [87].

To characterize these runaways, we considered the obsolescence rate $\Gamma = 1/\tau_0$. Figure 6.6b shows that Γ decreases with K and the $\Gamma(K)$ dependence is captured by the empirical relation

$$\Gamma = \Gamma_0 - G \ln K, \tag{6.5}$$

where the parameter Γ_0 defines longevity of the ordinary (low-cited) papers. Indeed, for small $K \sim 1$, Eq. 6.5 yields $\tau_0 = \Gamma_0^{-1} = 4.6, 9$, and 11.8 years for Physics, Economics, and Mathematics, correspondingly. Since citation trajectory of the ordinary papers is nearly exponential, it comes to saturation after $3\tau_0$. The long citation lifetime of the Economics and Mathematics papers, as compared to that of

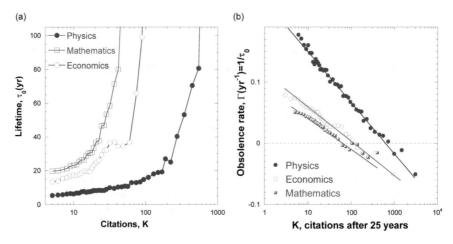

Fig. 6.6 (a) Citation lifetime τ_0 versus K, the number of citations after 25 years. τ_0 grows with increasing K and diverges at some K_r. The highly-cited papers with $K > K_r$ are runaways. The solid lines are the guide to the eye. (**b**) Obsolescence rate, $\Gamma = \tau_0^{-1}$. For each discipline, Γ decreases (lifetime increases) with the number of citations. Above certain K_r (760 for Physics, 113 for Economics, and 55 for Mathematics) the obsolescence rate Γ changes sign indicating the onset of the runaway behavior. Solid lines show empirical logarithmical dependence given by Eq. 6.5. Adapted with permission from Golosovsky [67]. Copyright (2017) by the American Physical Society. https://journals.aps.org/pre/abstract/10.1103/PhysRevE.96.032306

the Physics papers, is related to the propensity of these fields to cite old papers and to the more rapid growth of the number of publications in these fields.

Equation 6.5 indicates that the function $\Gamma(K)$ changes sign and becomes negative at certain $K_r = e^{\frac{\Gamma_0}{G}}$. Negative obsolescence rate indicates exponentially increasing number of citations—the runaway behavior. Thus, the papers with $K < K_r$ have finite lifetime and eventually become obsolete, the papers with $K > K_r$ are immortal—their citation career continues indefinitely. We found that the runaway threshold for the publication year 1984 is 760 citations for Physics, 113 citations for Economics, and 55 citations for Mathematics.

K_r, the threshold of runaways is determined by the parameter G defined in Eq. 6.5. Figure 6.4 shows that G is directly related to the nonlinear coefficient b in Eq. 6.2. Thus, decreasing $\Gamma(K)$ and increasing $c_{t,t-1}(K)$ dependences (Fig. 5.6) provide direct evidence of the nonlinear citation dynamics.

6.3 Universality of Citation Distributions

Radicchi et al. [141] were the first to notice that citation distributions for different disciplines and different publication years look very much the same. They conjectured that after dividing citation distribution by its mean, all these distributions collapse onto one universal curve. This conjecture was supported by their measurements. Subsequent studies [17, 38] extended this conjecture to the sets of publications from different journals, and different institutions. Subsequent careful measurements from the same group revealed some deviations from the one-parameter scaling [174] and the two-parameter scaling was suggested instead [139, 140]. In what follows, we examine the conjecture of universality basing on understanding of citation dynamics achieved in previous chapters.

In order to find out to which extent citation dynamics of scientific papers are universal, namely, are qualitatively the same for different disciplines, we use continuous approximation of our model and focus on a paper j with fitness η_j. To analyze citation dynamic of this paper, we divide its annual citation rate $k_j(t)$ by $m(t)$, the mean annual citation rate for all papers in this discipline published in the same year (see also Ref. [18].) $m(t)$ is related to the age distribution of references $r(t)$ through duality relation, $m(t) = e^{(\alpha+\beta)t} R(t)$ where $R(t) = R_0 r(t)$. We go to continuous approximation of the model, substitute these relations into Eq. 3.27 and find

$$\frac{k_j(t)}{m(t)} = \eta_j \left[\frac{A(t)}{r(t)} + q_j \int_0^t \frac{A(\tau)}{r(\tau)} e^{-(\tilde{\gamma}-q_j-\alpha-\beta)(t-\tau)} d\tau \right]. \tag{6.6}$$

The first term in the square brackets accounts for direct citations and the second term captures indirect citations. Here, $A(t)$ is the aging function for references. Since $A(t)$ and $r(t)$ are universal, dynamics of direct citations should be universal as well. With respect to indirect citations, we note that $A(t)$, $r(t)$, and $\tilde{\gamma}$ are nearly universal, while the growth exponents α, β and the parameter q are not. Indeed, $q \propto M P_0$, where M is the average number of citations and $P_0 = a(1 + b \ln K_j)$. Since $M \propto R_0$ and $a \propto 1/R_0$ (see Fig. 6.4), their product does not depend on R_0 and thus $q \propto (1 + b \ln K_j)$ where b still depends on R_0. Thus, in the expression for indirect citations, the growth exponents α, β and the nonlinear coefficient b are not universal. So far, as the citation dynamic of the paper is sensitive to α, β, and b, it is non-universal. Inspection of Eq. 6.6 shows that since α, $\beta << \tilde{\gamma}$, the most important parameter that depends on the discipline is b. Thus, citation dynamics of papers long after publication is not universal, as it is governed by indirect citations.

However, the early citation dynamics of papers, which is mostly determined by the first addend in Eq. 6.6, should be universal. Indeed, Fig. 6.7 shows that early citation distributions for different disciplines scale fairly well. This scaling indicates, that, in addition to universality of the early citation dynamics, the fitness distributions for different disciplines is the same. Indeed, we found earlier that the fitness distributions for Physics, Economics, and Mathematics follow the log-normal distribution with the same mean and variance, $\mu \sim -1.48$, $\sigma \approx 1.1$. Notably, US patent citation distribution has more or less the same $\sigma = 1.1$ [42]. The parameter σ gauges the spread of fitnesses, its universality means that the relative proportion of the low and highly-cited papers for each discipline is actually the same. This, probably, results from the same hierarchical and organizational structure of the research teams and institutions for all disciplines. We do not know what is so special

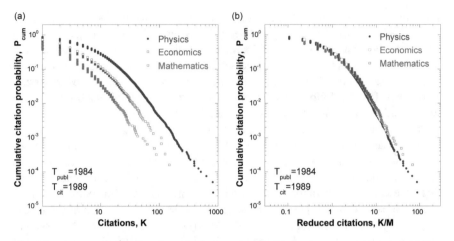

Fig. 6.7 Annual cumulative citation distributions for 40,195 Physics papers, 6313 pure Mathematics papers, and 3043 Economics papers—all published in 1984. Citations are counted at the end of 1989. (**a**) Cumulative citations K. (**b**) Reduced cumulative citations K/M, where M is the mean of the corresponding distribution. All three curves collapse onto one universal dependence

in the log-normal distribution with $\sigma \sim 1$, but it is one of the narrowest log-normal distributions observed in science [102].

6.4   Uncited Papers

The big question related to the phenomenon of uncitedness is whether it is an inevitable companion or a burden on science. The essay of van Noorden [168] summarizes this issue fairly well. Here, we address this question in a quantitative way and consider how the fraction of uncited papers depends on time and whether this dependence is captured by our model or not. Figure 6.8 shows $f_{uncited}(t)$ for three disciplines. Our model captures these dependences fairly well (see Fig. 5.9).

Our model yields a closed expression for $f_{uncited}(t)$. Indeed, Eq. 6.1 indicates that this function can be found from the number of non-events in the Poisson process, $f_{uncited}(\eta, t) = e^{-\Lambda}$, where $\Lambda = \eta R_0 \int_0^t \tilde{A}(\tau)d\tau$ is the Poisson rate for citations garnered by a paper during time t. Following this logic, for a set of papers with different fitnesses, all published in the same year, the fraction of uncited papers is

$$f_{uncited}(t) = \int_o^\infty e^{-\eta R_0 \int_0^t \tilde{A}(\tau)d\tau} \rho(\eta)d\eta, \qquad (6.7)$$

where $\tilde{A}(t)$ is the aging function for citations, $\rho(\eta)$ is the fitness distribution, and R_0 is the average reference list length. Of these parameters, $\rho(\eta)$ is universal while R_0

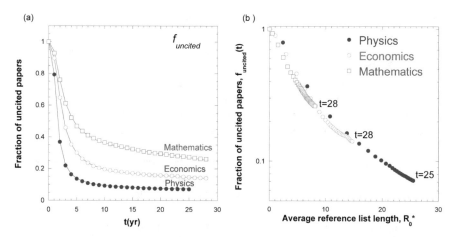

Fig. 6.8 $f_{uncited}$, the fraction of uncited papers. (**a**) Raw data. (**b**) The fraction of uncited papers versus $R_0^*(t)$ (see Eqs. 6.8 and 6.9)

and $\tilde{A}(t)$ are specific for each discipline. Notably, these two do not appear in Eq. 6.7 separately. It contains their product,

$$R_0^*(t) = R_0 \int_0^t \tilde{A}(\tau)d\tau. \tag{6.8}$$

Basing on Eq. 6.8, we recast $f_{uncited}(t, R_0)$ as a function of only one parameter $R_0^*(t)$, in such a way that Eq. 6.7 transforms to

$$f_{uncited}(R_0^*) = \int_o^\infty e^{-\eta R_0^*}\rho(\eta)d\eta. \tag{6.9}$$

Figure 6.8b shows three $f_{uncited}(t, R_0)$ dependences, shown on Fig. 6.8a, versus R_0^*. All three $f_{uncited}(R_0^*)$ dependences collapse onto a single curve suggested by Eq. 6.9. On the one hand, this is another indication of the universality of the fitness distribution. On another hand, since Eq. 6.9 does not contain free parameters, its good correspondence with the data is another proof of the validity of the model. Figure 6.8b is an important advance over past attempts to relate the uncitedness to the mean number of citations [27, 80]. Here, we show that the proper parameter which underlies scaling of $f_{uncited}(t)$ for different disciplines is not $m(t)$ but R_0^*. Although these two functions are closely related, they are not identical.

In what follows, we further analyze the $f_{uncited}(R_0^*)$ dependence shown in Fig. 6.8b. It captures not only the time dependence of $f_{uncited}$ but its dependence on the reference list length as well. If we consider it as $f_{uncited}(R_0)$ dependence, we learn that $f_{uncited}$ decreases with increasing R_0. Equation 6.9 shows this very well. Indeed, it yields $f_{uncited} = 1$ for $R_0 = 0$. This is obvious, since if papers do not cite one another, all of them remain uncited. On another hand, for $R_0 \to \infty$, Eq. 6.8 yields $f_{uncited} \to 0$. This is also reasonable, because if the reference lists of papers are very long, all papers will be eventually cited. These extreme cases are captured by Eq. 6.9. With respect to the actual values of $f_{uncited}$ in the long time limit of 25–28 years after publication, Fig. 6.8a yields 7.1%, 14%, and 26% for Physics, Economics, and Mathematics papers published in 1984. However, these numbers are not final and, as Fig. 6.8b shows, the fraction of uncited papers continuously decreases. The reason for this is not only the time after publication but slow increase in the average reference list length R_0 as well.

The very fact that Eqs. 6.7 and 6.9 capture the number of uncited papers is significant. Indeed, all independent parameters and variables in these equations are related to citations and their dynamics, none of them is related to non-citations. Non-citations here come as a byproduct of citations and this is the Poisson statistics which makes them inseparable [27]. All this indicates that the uncited papers constitute the inherent part of the scientific enterprise, namely, uncited papers are not unread [148, 168].

6.5 Summary

The input parameters of our model of citation dynamics are fitness distribution $\rho(\eta)$, aging functions for references and citations, $A(t)$ and $\tilde{A}(t)$, correspondingly; the probability kernel of indirect citations $P(K, t) = P_0 e^{-\gamma t}$, and the average reference list length R_0.

- The fitness distribution is universal. It is a log-normal distribution with the mean $\mu \approx -1.48$ and standard deviation $\sigma \approx 1.1$.
- The aging function for references is universal, while the aging function for citations is not and it is specific for each discipline. The difference between disciplines is related to their different growth rates.
- The exponent of the kernel of indirect citations is universal, $\gamma \approx 1.2 \, \text{yr}^{-1}$, but the probability amplitude P_0 is not. This probability depends logarithmically on the number of citations K and this is the $P_0(K)$ dependence which makes citation dynamics nonlinear. The slope of this dependence is non-universal and depends on R_0 in a non-trivial way.
- The most important parameter that makes citation dynamics so different for different disciplines, is R_0—the average reference list length.

If the probability amplitude of indirect citation were constant, $P_0 = const$, citation dynamics becomes linear and citation dynamics for different disciplines would differ only in scale set by R_0. However, since the probability amplitude of indirect citation, P_0, depends on R_0 in a non-trivial way, such simple scaling does not work. Hence, although the difference in citation dynamics for different disciplines reduces to a single parameter R_0, the scaling does not work and citation dynamics is not universal.

Chapter 7
Prediction of Citation Dynamics of Individual Papers

Abstract We apply stochastic model of citation dynamics of individual papers developed in Chap. 3 to forecast citation career of individual papers. We focus not only on the estimate of the future citations of a paper but on the probabilistic margins of such estimate as well.

Keywords Citation forecast · Fitness · Timeliness · Citation trajectory

7.1 Introduction

The interest in predicting citation behavior of scientific papers is motivated by the need to forecast the journal impact factor, early identification of the breakthrough papers, career considerations, etc. (see Ref. [40, 164, 185]). Prediction is usually based on a priori and a posteriori factors that, in principle, can determine citation career of a paper. The former factors are set at the moment of publication and these are subject, title, author's previous record and reputation [34, 85, 132, 159, 181, 182], venue (journal) [46, 93], the length and the composition of the reference list [46, 86, 167, 181], the style of the paper [46, 55, 98], etc. The difficulty of this approach is that the most important attributes, such as novelty, originality, significance, timeliness of the results, etc. are qualitative and their quantification is challenging. A brilliant example of such quantification is Ref. [167] which managed to characterize the novelty of a paper through diversity (frequency of atypical combinations) of its references. A posteriori factors develop during short time after the paper has been published and these include the "impact factor"—the number of citations during a short period after publication [1, 31, 34, 100, 175], and the place that the paper occupies in its community. There are two complementary approaches to predict citation career of a paper basing on these factors.

Computer scientists focus more on a priori factors. They take a large set of papers whose citation career has been evolving for a long time and use it for training, namely, they measure correlation between these factors and the number of citations of a paper in the long time limit. Then, the factors are ranked according to their importance and predictive model is built by machine learning. The general

M. Golosovsky, *Citation Analysis and Dynamics of Citation Networks*,
SpringerBriefs in Complexity, https://doi.org/10.1007/978-3-030-28169-4_7

consensus is that predictive algorithm shall use several factors or combination of them [16, 159], whereas the relative weight of these factors for different disciplines can vary. It has been also realized that linear correlations do not tell the whole story [34, 72, 100, 175] and predictive algorithm shall be better nonlinear, similar to that of Ref. [100]. When the predictive algorithm has been validated, it works as follows. For a new paper, one determines all relevant factors and builds a prediction. The result of prediction is the number of citations of a paper after some predetermined time. Although this prediction is probabilistic, the margins of predictability were never studied properly.

The approach of researchers with the background in natural sciences is different. They focus more on a posteriori factors, such as recent citation history of a paper. They construct empirical models of citation dynamics which are based on some predetermined scheme of the citation process, namely, they assume a certain strategy that the author of a new paper adopts when he cites the previous studies. This model predicts a future citation behavior of a paper basing on its citation history and several paper-specific parameters, the most important of them being fitness, a hidden parameter that can be reliably estimated only after citation career of the paper has been developing for 2–3 years [74]. When the model has been constructed and validated, the prediction is performed as follows. One takes a new paper and, by studying its initial citation history, makes a probabilistic estimate of its fitness and other specific parameters. After such estimate has been made and the corresponding parameters have been substituted into the model of citation dynamics, it predicts the number of citations of this paper in the long time limit. This approach has been most completely embodied in the Wang-Song-Barabasi model [175].

Bibliometric analysis considers both a priori and a posteriori factors. The researchers in this area have long recognized that the early citation history of a paper is a good predictor of its future success. On another hand, they were the first to draw attention to sleeping beauties [64, 84, 169, 175], the papers that started to gain popularity long after publication. Many important papers exhibited the sleeping beauty behavior which no model of citation dynamics can predict. Thus, the presence of such papers sets a limit to prediction of the future citation count of a paper. On another hand, this poor predictability is what makes science fun for so many researchers.

7.1.1   Our Goal

We use our stochastic model of citation dynamics developed in Chap. 3 to forecast the future citation career of a paper. Our model includes several empirical parameters, some of them are common to the whole discipline while all individual attributes of the paper are lumped into one parameter—fitness which does not vary with time. Our first goal is to explore the limits of predictability of the citation career of a paper with a given fitness, the uncertainty of prediction being related to intrinsic stochasticity of the citation process. Our second goal is to quantify the ingredients of fitness, in particular, we show how one can quantify such attribute of a paper as timeliness of results.

7.2   Probabilistic Character of the Citation Process and Its Implications with Respect to Predictability of Future Citations

Citation process is stochastic, the stochasticity imposes limits on the predictability of future citations. Moreover, as we showed in Chap. 5, citation dynamic of a paper follows a self-exciting (Hawkes) process whereby past fluctuations are amplified. The positive feedback between past fluctuations and future citations renders the task of long-term prediction of citation behavior of a paper almost futile and limits predictive algorithms to the range of 2–3 years. In what follows, we illustrate this by measurements.

7.2.1   Divergence of Citation Dynamics of Similar Papers: Measurements

To explore the limits of predictability of citation count basing on early citation history of a paper, we considered sets of all Physics papers published in 1984 that garnered a certain number of citations during first 2–3 years after publication. As an example, Fig. 7.1 displays citation dynamics for the set of papers that garnered 30–31 citations during 1984–1986. Although these 89 papers represent an extremely homogeneous set (same discipline, same publication year, almost the same citation prehistory), the divergence in their citation trajectories is striking. We ranked the papers according to the number of citations garnered 25 years after publication. The list opens with the pair of papers: "Lower critical dimension of the random-field Ising model—a Monte-Carlo study" by D. Andelman, H. Orland, and L.C.R. Wijwardhana, Phys. Rev. Lett. (40 citations); and "Fractional quantum Hall-effect at filling factors up to $\nu = 3$" by G. Ebert, K. von Klitzing, and J.C. Maan, J. of Physics C—Solid State (49 citations). The lists ends with the papers: "Dynamics of supercooled liquids and the glass transition" by U. Bengtzelius, W. Gotze and A. Sjolander, J. of Physics C—Solid State (798 citations); and "Embedded-atom method: Derivation and application to impurities, surfaces, and other defects in metals" by M.S. Daw and M.I. Baskes, Phys. Rev. B (2254 citations). All four papers were published actually in the same journals in the same year and they have the same citation prehistory. The two former papers are important works with typical citation dynamics. The two latter papers exhibit strongly different citation dynamics and the last one is a runaway. This is another evidence of the highly individual citation dynamics of citation classics which was extensively studied by Redner [144].

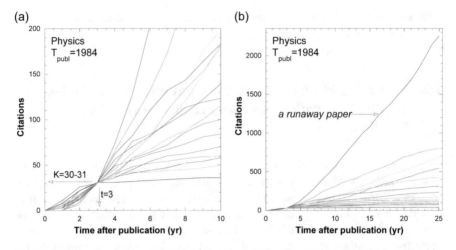

Fig. 7.1 Citation dynamics of 89 Physics papers published in 1984 which garnered 30 or 31 citations by 1986 (2–3 years after publication). Although the initial citation trajectories of these papers are very similar, they quickly diverge, in such a way that after 25 years, the number of their citations varies between 40 and 2254. (**a**) A short time span after publication. (**b**) A long time span after publication. Note a runaway paper whose citation dynamics does not come to saturation in 25 years. Adapted with kind permission of European Physical Journal (EPJ) from Golosovsky and Solomon [69]. Copyright by the Springer-Verlag 2012

Figure 7.1a shows, that if we use the early citation history of a paper as a basis for prediction, the time span for a reasonable prediction is only 2–3 years. Indeed, if citation distribution of some set of papers at $t = 3$ is a delta-function, citation distribution of the same set at $t = 5$ is so wide that its width is comparable to the mean. This divergence is partially due to the spread of fitnesses in this set of papers. However, the most important source of this divergence is the internal stochasticity of citation process. In what follows, we focus on it.

7.2.2　Divergence of Citation Dynamics of the Papers with the Same Fitness: Numerical Simulation

We explore here the following question: if we had known paper's fitness—what are the margins of predictability of its citation trajectory? To answer this question, we analyzed the relation between the paper's fitness and the number of citations it garners in the long-time limit. This was done using our calibrated and verified model of citation dynamics presented in Chap. 3. We wish to estimate, $K^{\infty}(\eta)$, the expected number of citations after 25 years for the paper with a certain fitness η. To this end, we performed numerical simulations based on Eq. 3.20 with parameters for Physics papers published in 1984. We considered 4000 papers with the same fitness η, found statistical distribution of their citations after 25 years, and measured

(a)

(b)

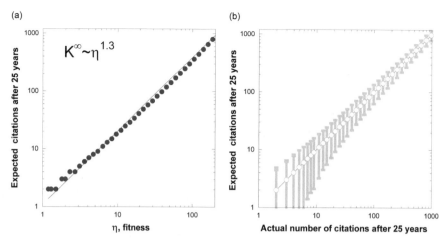

Fig. 7.2 (**a**) Expected number of citations after 25 years (the long time limit of the number of citations), $K^{\infty}(\eta)$, in dependence of the paper's fitness η. Numerical simulation for 4000 papers with the same η. The simulation is based on Eq. 3.20 with the parameters for Physics papers published in 1984. Continuous line shows an empirical power-law dependence, $K^{\infty} \propto \eta^{1.3}$. (**b**) Actual number of citations after 25 years versus expected number. The error bars show the width of the distribution, $K^{\infty} \pm std(K^{\infty})$. Citation distributions are broad for low η and narrow for high η

the mean K^{∞} and the width of this distribution. We consider K^{∞} as the expected number of citations in the long time limit. This was repeated for several values of fitness.

Figure 7.2a shows that the expected number of citations, K^{∞}, grows nonlinearly with fitness η. Figure 7.2b focuses on the width of the K^{∞} distribution. We observe that for the papers with low η, citation distribution in the long time limit is wide, while for the papers with high η, citation distribution in the long time limit is narrow. This means that while citation dynamic of a low-fitness paper strongly depends on chance, citation dynamics of the high-fitness papers are more deterministic.

In particular, Fig. 7.2b shows that if expected number of citations in the long-time limit is 3, the actual number of citations can be anything between 0 and 7; if expected number of citations is 10, the actual number can be between 3 and 20, if the expected number is 100, the actual number can be between 50 and 130, if the expected number is 1000, the actual number can be between 700 and 1200. If we compare two papers that garnered 3 and 20 citations in the long-time limit, they can have the same fitness η, namely, they are most probably in the same "quality" league. Two papers that garnered 700 and 1200 citations are probably in the same "quality" league, namely they can have the same fitness. But the papers that garnered 100 and 1000 citations should have different fitness and belong to different "quality" leagues.

Figure 7.3 shows K^{∞} distributions from a slightly different perspective. We observe that a paper which is *worth* 10 citations, with 10% probability can garner less than 3 or more than 18 citations; a paper which is *worth* 100 citations, with 10% probability can garner less than 54 or more than 135 citations; a paper which

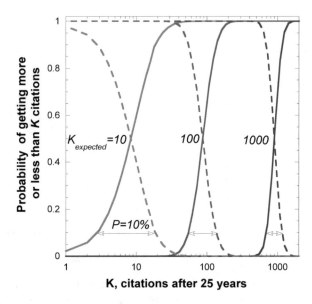

Fig. 7.3 Statistical distribution of the number of citations after 25 years for the papers with the same fitness η. Numerical simulation based on Eq. 3.20 for 4000 papers. The mean of the distribution, K^∞, is indicated at each curve. Continuous lines show cumulative probability of getting more than K citations, $\int_K^\infty p(K)dK|_{\eta=const}$. Dashed lines show complementary probability of getting less than K citations, $\int_0^K p(K)dK|_{\eta=const}$. The intervals of the 10% probabilities of having $K_{expected}$ citations are shown by arrows

is *worth* 1000 citations, with 10% probability can garner less than 680 or more than 1150 citations.

It should be noted that our model assumes constant fitness through the whole citation career of the paper. Similar assumption was adopted by Ref. [87] in their description of Web-pages popularity and it was justified by measurements. This assumption is reasonable for ordinary papers but not for sleeping beauties, that can be dormant for a long time and then become popular.

7.2.3 Fitness Estimation

Refs. [160, 175] associate fitness with the ultimate impact of the paper, namely the number of total citations in the long-time limit; Ref. [54] determines paper fitness by ranking; Ref. [75] estimate patent fitness as a combination of attributes found through factor analysis, Ref. [154] associates fitness with the number of citations during a couple of years after publication. We define fitness slightly differently, namely, η is the number of direct citations in the long-time limit. Obviously, this definition cannot be a basis for prediction of future citations since it can be used only when citation career of a paper is close to completion.

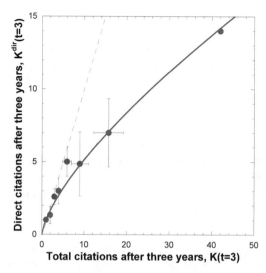

Fig. 7.4 The relation between direct citations and total citations during first 3 years after publication for Physics papers published in 1984. Publication year corresponds to $t = 1$. The dashed line shows linear approximation, $K^{dir}(3) = K(3)$. This approximation is good only for low-cited papers. Red continuous line shows the empirical approximation $K^{dir}(3) = [K(3)]^{0.7}$ which shall be used for highly-cited papers. The fitness is estimated from Eq. 3.20 as $\eta_j = \dfrac{K_j^{dir}(3)}{R_0 \sum_1^3 \bar{A}(t)}$

To work out operational definition of fitness that can be used for prediction, we note that the fitness in our sense is related to initial citation rate, since at the beginning of the citation career of a paper, citations are predominantly direct. We base our operational definition of fitness on the "magic of three years", well-recognized in bibliometrics. Namely, the number of citations garnered by a paper during first 2–3 years after publication (for computer science papers this initial period is 0.5–1 year) is a basis for fitness estimation. Figure 7.4 shows that the relation is nonlinear (see also Fig. 7.2a) and from this calibration plot we estimate fitness.

7.3   Fitness Estimation Basing on Paper's Content

We believe that η_i is determined by the journal (venue), the number of researchers in the area, reputation of the research group, and last but not least—by the paper's novelty, timeliness, and quality although the latter can be subjective notion. It should be noted, however, that the paper's fitness and the number of citations gauge not the quality of a paper but its impact. Note, that even erroneous paper can have a great impact. On the other hand, the impact can depend on the factors unrelated to paper's content—institution, reputation of the research group, catchy title, etc.

For example, Brot and Louzoun showed [23] that the name of the first author matters for citation count, in particular, the Physics papers whose first author's name starts from the letters A, B, C, in the long time limit have ∼10% more citations than the papers whose list of authors starts from X, Y, Z. (May be, success of the famous Alpher-Bethe-Gamow paper partially derives from the lucky combination of their last names?)

To further demonstrate the importance of the author's name for success of the paper, we consider anectodal evidence based on a couple of papers. Indeed, Richard Lewontin and Jack Hubby made a landmark study in molecular evolution while collaborating in the University of Chicago. To get equal credit for their contribution, their scientific report was published as two companion papers with very similar titles and subjects:

1. J.L. Hubby and R.C. Lewontin, "A molecular approach to study of genic heterozygosity in natural populations. **1**. Number of alleles at different loci in drosophila pseudoobscura", Genetics 54(2), 577–594 (1966).
2. R.C. Lewontin and J.L. Hubby, "A molecular approach to study of genic heterozygosity in natural populations. **2**. Amount of variation and degree of heterozygocity in natural populations of drosophila pseudoobscura", Genetics 54(2), 595–609 (1966).

The main difference between these two papers is the order of authors. By 2018, the second paper got around 900 citations while the first paper got only around 500 citations! This difference is explained by the fact that, when the papers were first published in 1966, Lewontin, who was 3 years older than Hubby, was better known in the scientific community. Thus, researchers preferred to cite the paper in which Lewontin was the first author. Eventually, Hubby became also well-known, the paper in which he was the first author got fair credit and a large number of citations. However, citation count of the Lewontin paper remained bigger due to impressive head start. On another hand, do citation counts of these two papers reflect difference in their "quality"? Our model and Fig. 7.2 show, that the probability of two papers, which garnered 500 and 900 citations in the long time limit, to have the same fitness, is ∼10%. This probability is not small, hence it is quite probable that the papers of Lewontin and Hubby are in the same "quality" league.

7.4 Timeliness of Results

One of the important criteria, which the editor and reviewers use in their eavluation of submitted papers, is the timeliness of results. This criterion singles out the papers that deal with a hot topic. In our parlance, the paper that focuses on hot topic has enhanced fitness as compared to the paper belonging to the mature research direction. How one can quantify the corresponding contribution to fitness?

Suppose that at year t_0 there appeared one or several breakthrough papers which were followed by a flurry of subsequent developments. This means that a new field (hot topic) has been born. The number of publications in this new field starts to grow explosively and then saturates. As we have shown in Chap. 2, the authors are conservative in their citing habits, and the length and the age composition of their reference lists remains more or less the same. In particular, the papers that were published in the same year constitute $\sim$2–3% of the reference list, the papers published an year before constitute $\sim$8–10%, the papers that were published 2 years before also constitute $\sim$8–10%, etc. Thus, the papers that were published long after the onset of a new topic have big choice in choosing their references, while the papers published soon after the onset of a new topic have a very limited choice for filling their reference list and all choose the papers that were published close to the onset. Thus, the papers that were published soon after the birth of a new field, namely, timely papers, shall have enhanced number of citations (enhanced fitness).

To put these considerations into quantitative terms, we consider a new field that appeared at time t_0. We denote the annual number of publications in this field by $N(t_0 + t)$. Equation 3.20 yields the average number of direct citations that the paper in this field, which was published in year $t_0 + t$, garners during three subsequent years,

$$K^{dir}(t_0 + t, t_0 + t + 3) = \bar{\eta}(t_0 + t)R_0(t_0 + t)\sum_1^3 \tilde{A}(\tau)N(t_0 + t + \tau). \quad (7.1)$$

where $\bar{\eta}$ is the average fitness of the papers in the new field which were published in year $t_0 + t$, R_0 is the average reference list length of the papers published in year t_0, $\tilde{A}(\tau)$ is the aging function for citations. Note also, that $\tilde{A}(\tau) = A(\tau)e^{(\alpha+\beta)\tau}$ where $A(\tau)$ is the aging function for references. The reference-citation duality yields average fitness for the papers published in year $t_0 + t$,

$$\bar{\eta}(t_0 + t) = \frac{\sum_1^3 A(\tau)\frac{N(t_0+t+\tau)}{N(t_0+t)}e^{\beta\tau}}{\sum_1^3 A(\tau)e^{(\alpha+\beta)\tau}}. \quad (7.2)$$

If the new field grows with the same rate as the whole discipline, namely, $\frac{N(t_0+t+\tau)}{N(t_0+t)} = e^{\alpha\tau}$, then $\bar{\eta}$ does not depend on t. However, if this new field grows faster than the whole discipline, then $\bar{\eta}$ is enhanced.

Figure 7.5 illustrates these considerations. We know that a hot topic usually appears abruptly and can be identified through a burst of citations and publications [99]. We choose several such research areas in Physics with well-defined onset t_0, with some of these areas the author of this book has had personal experience. Using Web of Science, we found all papers belonging to each of these topics, that were published in year $t_0 + t$. For each t, we measured annual number of papers and statistical distribution of the number of citations garnered by them during first 3 years after publication. Then we determined the mean and the width of these distributions. Using Eq. 7.2 and Fig. 7.4, we found the average fitness of the papers in each topic published in year $t_0 + t$, basing on the mean of the distribution. On another hand, we estimated this fitness using Eq. 7.2. Figure 7.5 shows that the model prediction based on Eq. 7.2 captures our measurements perfectly well. (The fitness here differs from what we introduced in Chap. 3 by the factor R_0, the average reference list length.)

Figure 7.5 implies that any paper published soon after the new topic appeared, has a good head start and this quantifies the "first mover advantage" introduced by Newman [124]. However, this does not mean that the papers published long after the onset of a hot topic doomed to be undercited. In fact, Fig. 7.5 shows only the mean of the fitness distribution for each year. The actual fitness distribution is very wide and its width is comparable to the mean. Hence, at each moment after the onset of a hot topic there are many papers whose fitness considerably exceeds the average one.

7.5   Summary

In this chapter we demonstrated that our stochastic model of citation dynamics can be a basis for predicting citation trajectory of papers. This model shall be compared to the physics-inspired predictive model developed by Wang et al. [175]. Pham et al. [134, 135] developed a software package based on this model and demonstrated that it is a valid predictive tool. This model includes three paper-specific parameters: fitness η, immediacy μ, and σ. To determine these parameters, one needs to measure initial citation trajectory of a paper, 2–3 years are not enough. As a predictive tool, this model works best for the highly-cited papers. Although this deterministic model predicts citation trajectory of a paper, it cannot specify probabilistic margins of the prediction. On the contrary, our probabilistic model includes only one paper-specific parameter—fitness, it does provide probabilistic margins of the future citation count. However, our model works better with ordinary papers and does not predict well citation trajectories of the highly-cited papers. Thus, our model is complementary to that of Ref. [175].

What are its possible applications? We believe that our model can be used for forecasting the 5-year journal impact factor. The papers published in the same year in one journal represent more or less homogeneous set of papers, hence predicting the mean number of citations for this set is more reliable than predicting citation trajectory of a single paper. On another hand, our model can give probabilistic margins of such prediction.

Another application can be the early identification of the breakthrough papers. So far, this was done by analyzing diversity and age structure of the reference list of papers [119, 167], diversity and interdisciplinarity of paper's content [137], or through identification of the atypical citation trajectory, corresponding to sleeping beauties [84]. An important question is how soon can we identify such rising star? Obviously, if the paper (or patent) gets more citations than what is expected from the ordinary paper published in the same year and in the same journal, then this is a

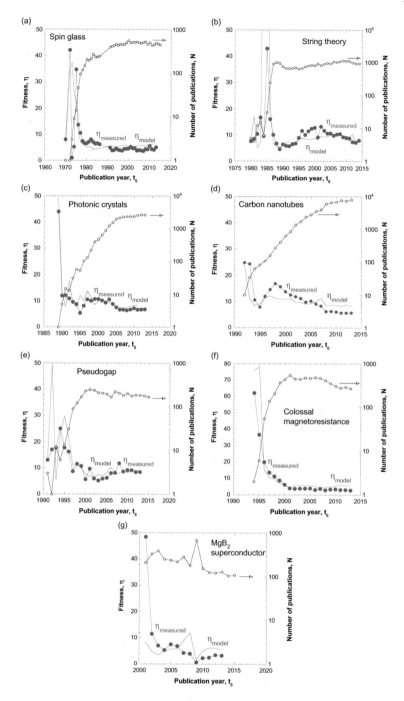

Fig. 7.5 Paper's fitness for several Physics research topics. Open black circles show number of research articles for each topic. Filled circles show our measurements of fitness based on the number of citations garnered during first three years after publication. Blue continuous lines show model prediction based on Eq. 7.2. (**a**) Spin glass. (**b**) String theory. (**c**) Photonic crystals. This was an active research field initiated by Eli Yablonovich and Sajeev John in 1991. Eventually, this field broadened and evolved into metamaterials, in such a way that the term "photonic crystal" faded out, although the field of photonic crystals continues to flourish. (**d**) Carbon nanotubes. (**e**) Pseudogap. (**f**) Colossal magnetoresistance. (**g**) MgB$_2$ superconductor

candidate to be a breakthrough paper [105]. On another hand, the deviation from the ordinary citation trajectory may be accidental. Our model can make an estimate of the probability of the enhanced citation count in order to judge whether it occurred by chance or not.

Chapter 8
Power-Law Citation Distributions are Not Scale-Free

Abstract We analyze time evolution of statistical distributions of citations to scientific papers published in the same year. While these distributions seem to follow the power-law dependence, we find that they are nonstationary and the exponent of the power-law fit decreases with time and does not come to saturation. We attribute the nonstationarity of citation distributions to different longevity of the low-cited and highly-cited papers. By measuring citation trajectories of papers, we found that citation careers of the low-cited papers come to saturation after 10–15 years while those of the highly-cited papers continue to increase indefinitely. When the number of citations of a paper exceeds some citation threshold, it becomes a runaway. Thus, we show that although citation distribution can look as a power-law dependence, it is not scale-free and there is a hidden dynamic scale associated with the onset of runaways. We show that our model of citation dynamics based on copying/redirection/triadic closure accounts for these issues fairly well.

Keywords Power law distribution · Scale-free distribution · Citation lifetime

8.1 Introduction

8.1.1 Power-Law Distributions

Highly-skewed statistical distributions were discovered more than a century ago and remain an object of intense research (see Refs. [41, 114, 123, 136] for comprehensive reviews). The most important among these fat-tailed continuous distributions are the power-law

$$p(x) \propto x^{-\nu}; \ x \geq x_{min} \tag{8.1}$$

and the shifted power-law (Pareto II) distribution

$$p(x) \propto (x + w)^{-\nu}; \ x \geq x_{min}, \tag{8.2}$$

© The Author(s), under exclusive license to Springer Nature Switzerland AG 2019
M. Golosovsky, *Citation Analysis and Dynamics of Citation Networks*,
SpringerBriefs in Complexity, https://doi.org/10.1007/978-3-030-28169-4_8

where $p(x)$ is the probability density function, ν is the exponent, and w is the shift. In contrast to the Gaussian distribution with its finite moments, the moments of the power-law distributions can diverge. In particular, the mean diverges for $\nu \leq 2$ and the variance diverges for $\nu \leq 3$. Thus, the first and the second moments of the power-law distribution with $2 \leq \nu \leq 3$ are determined by its tail and this is the reason why these distribution are named fat- or heavy-tailed.

The discrete analogue of Eq. 8.2 is frequently represented by the Waring distribution [26, 63, 111] which is also known as the shifted Yule-Simon distribution [41, 114, 123],

$$p(k) = \frac{B(k + w, \nu)}{B(w, \nu - 1)}, \tag{8.3}$$

where $B(a, b) = \frac{\Gamma(a)\Gamma(b)}{\Gamma(a+b)}$ is the Euler beta-function. The parameter ν is the analogue of the exponent and w is the shift. The mean of the Waring distribution is

$$M = \frac{w}{\nu - 2} \tag{8.4}$$

and it diverges for $\nu \leq 2$. For $k \gg w$, Eq. 8.3 reduces to the Zipf-Mandelbrot distribution,

$$p(k) \approx (\nu - 1)\frac{w^{\nu-1}}{(k + w)^{\nu}}, \tag{8.5}$$

which is the discrete analogue of Eq. 8.2.

Another class of the highly-skewed distributions is the log-normal,

$$p(x)dx = \frac{1}{\sigma\sqrt{2\pi}}e^{-\frac{(\ln x - \mu)^2}{2\sigma^2}}\,d\ln x, \tag{8.6}$$

where μ and σ characterize, correspondingly, the mean and the variance.

A peculiar property of the power-law and log-normal distributions is that they are scale-free, namely, the probability density functions $p(x)$ and $p(x/S)$, where S is a constant, have the same shape and are just shifted on the log-log scale (this is also true for Eq. 8.3 when $k \gg w$), in other words, these distributions or at least their tails are self-similar. The scale-free property of the highly-skewed distributions has been a source of fascination for many physicists that sought deep analogies with other scale-free phenomena such as fractals, phase transitions and critical phenomena [7, 28, 48, 123, 158, 177].

8.1.2 Experimental Assessment of the Fat Tailed Distributions

The fat-tailed distributions are frequently invoked to account for degree distributions in complex networks, in particular, for characterization of citation distributions. The simplest way of doing this is to plot the measured degree distribution on the log-log scale. The straight line indicates the power-law while the parabola suggests the log-normal distribution. In practice, the test for curvature on the log-log scale doesn't discriminate well between these two distributions. Indeed, if the log-normal distribution is very wide, then a large piece of parabola looks like a straight line. For discrete distributions, the situation is even worse since the log-log plot of the discrete power-law distribution (Eq. 8.3) has convex shape at small degrees and this further exacerbates the problem of distinguishing this distribution from the log-normal. Even if the log-log plot of a statistical distribution or of a part of it looks like a straight line, to find its slope is not an easy task [165]. Indeed, since most distributions round up at small degrees, to measure the slope one shall cut off the small degrees and focus on the tail of the distribution. This cutoff procedure is subjective and is a source of uncertainty [41]. The difficulties of experimental identification of the power-law degree distribution generated a substantial controversy of whether degree distributions in complex networks are better described by the power-law, log-normal, or stretched exponential [22, 41, 114, 123].

The history of assessment of citation distributions is a good example of such controversy. Beginning from the works of de Solla Price [45], citations were fitted by a discrete power-law distribution with the exponent $\nu = 2.5$–3.16 [41, 142]. Afterwards, the fashion shifted towards log-normal or discretized log-normal fitting function with $\sigma = 1$–1.2 [51, 102, 144, 160, 165]. Recent studies [2, 24] claimed again the power-law distribution with the exponent ν varying between 3 and 4.

Why is it so important to find the functional form of degree distribution of a complex network? The motivation for this was based on the belief [115, 161] that this functional form is a clue to the mechanism of network growth. However, to pinpoint the growth mechanism from degree distribution has proved ineffective and ambiguous. Therefore, Mitzenmacher [115] suggested to leave attempts of derivation of the network generating mechanism from degree distributions and to measure this mechanism directly, for example, by applying the time-series analysis.

We closely followed this suggestion and determined the growth mechanism of citation network from microscopic measurements. This mechanism has been described in previous chapters of this book. In this chapter, we address the following question: what is the functional form of degree distribution in citation networks? We approached this question from two directions. First, we chose several well-defined citation networks. For each one of them we measured how its degree distributions evolves with time. We analyzed these citation distributions empirically using commonly accepted strategies and found that they are non-stationary and do not converge to some limiting distribution, even in the long time limit. This nonstationarity explains why previous efforts to characterize degree distribution in citation networks were so ambiguous. Second, we modeled these citation distributions using our stochastic model of citation dynamics and found explanation for the nonstationarity. It turns out, that nonstationarity of citation distributions and their "power-law" shape both originate in *nonlinear* citation dynamics. In particular, nonlinearity introduces a certain scale, in such a way that citation networks can no longer be considered scale-free.

8.2 Empirical Characterization of Citation Distributions

To find the functional form of citation distributions, we chose several well-defined research fields and focused on all original research papers (overviews excluded) published in the same year. In particular, we considered the fields of Physics, Mathematics, and Economics published in 1984. For each field, the fraction of papers that garnered K citations during t years after publication, yields the probability density function, $p(K, t)$. Figure 8.1 shows the corresponding cumulative citation distributions, $P(K, t)$. We fitted these distributions using discrete power-law (Eq. 8.3) and discretized log-normal function (Eq. 8.6). The former fit assumes a straight tail in the log-log plot, while the latter fit assumes a convex tail. Figure 8.1 shows that for small t (early after publication) both fits perform equally well, while for later years, the discrete power-law fit is better. Indeed, for the most representative set of Physics papers, the tail of the distribution is straight, as suggested by Eq. 8.3, rather than convex, as suggested by Eq. 8.6.

In what follows, we use the discrete power-law (Waring distribution, Eq. 8.3) to parameterize citation distributions. Figure 8.2 shows time dependence of the fitting parameters w and ν that capture, correspondingly, the shift and the slope. The shift w increases with time and comes to saturation after ~ 10 years. The exponent ν continuously decreases but does not come to saturation. This means that even after 25 years citation distributions are not stationary and their tails still develop.

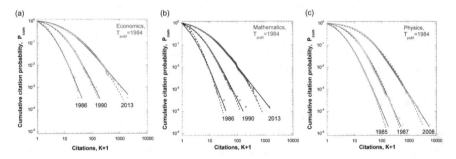

Fig. 8.1 Cumulative citation distributions for original research papers published in 1984. The points stay for measurements, the continuous lines show the discrete power-law fit (Eq. 8.3), the dashed lines show the log-normal fit (Eq. 8.6). (**a**) 3043 Economics papers, (**b**) 6313 Mathematics papers, (**c**) 40,195 Physics papers. Reprinted with permission from Golosovsky [67]. Copyright (2017) by the American Physical Society. https://journals.aps.org/pre/abstract/10.1103/PhysRevE. 96.032306

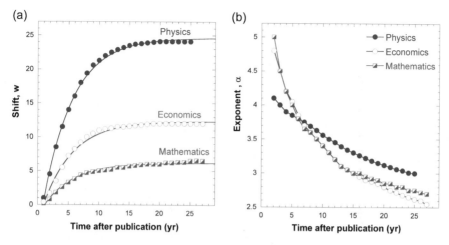

Fig. 8.2 Parameters of the discrete power-law fit as given by Eq. 8.3. (**a**) Shift, w. (**b**) Exponent, ν. Although the shift comes to saturation after 10 years, the exponent continuously decreases and does not come to saturation even after 25 years. Reprinted with permission from Golosovsky [67]. Copyright (2017) by the American Physical Society. https://journals.aps.org/pre/abstract/10.1103/PhysRevE.96.032306

Thus, contrary to common belief that citation distributions assume some limiting shape in the long-time limit, we found that citation distributions are nonstationary and do not achieve limiting shape even after 25 years. The reason for the nonstationarity is the highly-cited papers (see Fig. 8.3). As we show below, these papers eventually become runaways.

Indeed, the early citation distributions (Fig. 8.1, $t = 1$–5 year) can be fitted either by the log-normal or by the discrete power-law distribution with $\nu = 4$–5 (Eq. 8.3). Large exponent indicates that these distributions are "conventional", namely, their tails play only a minor role in defining the mean and the variance of the distribution. As time passes and papers garner more citations, citation distributions shift to the right. This shift comes to an end after ~10 years when citation career of the ordinary papers is over. Later on, citations are garnered mostly by the highly-cited papers which compose the tail of the distribution. The slope of the distribution becomes more gradual, since, as time passes, the tail moves fast to the right while the rest of the distribution slows down and eventually stays still. This is the reason why the power-law exponent ν decreases with time (Fig. 8.2). Eventually, the tail of the distribution comprises only runaway papers whose citation career prolongs indefinitely. While the tail continues to move to the right, the rest of the distribution stays still, in such a way that it never comes to saturation and eventually becomes concave.

So, why citation distributions with a concave tail are so rare? We claim that to observe a concave tail of the citation distribution requires a very long time window and a very large dataset that contains many runaway papers. Of all the sets presented in Fig. 8.1, even the largest set of 40,195 Physics papers is insufficient for this

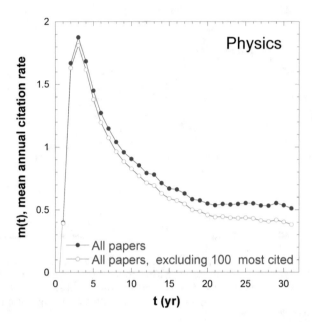

Fig. 8.3 Mean annual number of citations for all Physics papers published in 1984. The highly-cited papers which constitute the tail of the citation distribution, play disproportionately high role in the citation dynamics. Indeed, while 100 most cited papers make only 0.25% of the whole set, their contribution to the mean citation rate grows to 26% after 30 years

purpose. However, the 10-times bigger set of 418,438 Physics papers published in 1980–1989 does reveal the concave tail, as it is illustrated on Fig. 8.4.

8.3   Recursive Search Model Explains the Shape of Citation Distributions

8.3.1   Citation Distributions

As we already demonstrated in the previous chapter, our model successfully reproduces citation distributions for several disciplines. Here, we use this model for extrapolation to the time window inaccessible to measurements. Figure 8.5a shows that numerically simulated citation distributions do not become stationary and continue to develop even 50 years after publication. This explains why the empirical parameter—the exponent of the allegedly power-law tail of the citation distribution decreases with time and does not come to saturation (Fig. 8.2).

Which parameters of the model are responsible for the general shape of citation distributions? We already noticed that early citation distributions mimic the fitness distribution. For the Physics papers, the former can be represented either as log-

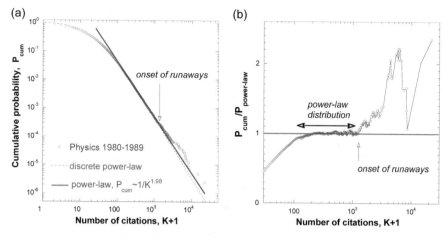

Fig. 8.4 (**a**) Citation distribution for 418,138 Physics papers published in 1980–1989 (overviews excluded) measured in 2008 [69]. Green dashed curve shows fit to discrete-power-law distribution (Waring distribution, Eq. 8.3). Red straight line shows a power-law fit, $P_{cum} \propto 1/K^{1.98}$. Both fits account fairly well for small and moderate K and fail for the tail, $K \geq 1200$. Note upward deviation of the tail of the measured distribution from both fits. This deviation indicates that the cumulative citation distribution is concave. (**b**) The ratio of the measured cumulative distribution to the power-law fit. While the deviation for $K < 100$ is trivial and is related to the rounding off of the distribution at small K, the upward deviation at $K \geq 1200$ indicates the runaway tail which contains more than 130 papers. Reprinted with permission from Golosovsky [67]. Copyright (2017) by the American Physical Society. https://journals.aps.org/pre/abstract/10.1103/PhysRevE.96.032306

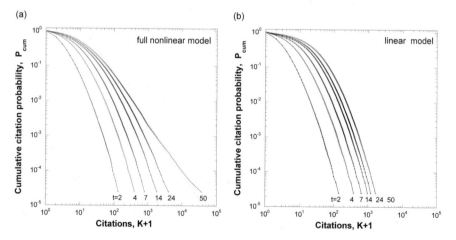

Fig. 8.5 Model prediction of citation distributions for 40,195 Physics papers published in 1984. t is the time after publication. (**a**) Numerical simulation based on Eq. 5.1 with the kernel given by Eq. 4.10 (full nonlinear model). The tails of all distributions besides the one for $t = 50$ are straight lines with the slope decreasing with time. However, the tail of the distribution for $t = 50$ has a concave shape. The concave tail indicates the presence of runaways. (**b**) Simulation based on Eq. 5.1 with a constant kernel, $P_0 = 0.54$ (a linear model). The tails of the simulated distributions are straight lines with time-independent slope

normal with $\mu = -1.48, \sigma = 1.12$ or Waring distribution with $\nu = 4$. Anyway, both fits result in convex distribution. Numerical simulation shows that, as time passes, citation distribution shifts to the right and its tail straightens, in such a way that it looks like a power-law dependence with $2 < \nu < 3$. This power-law-like dependence holds for 3–20 years after publication and this is the reason why citation distributions have been successfully fitted by the discrete power-law (Eq. 8.3). However, for longer time windows and for large sets of papers, the situation is different. Extrapolation of the simulation to $t = 50$ years shows that, in the long time limit, the distribution becomes concave (Fig. 8.5a, $t = 50$) rather than remaining straight. This means that the straight tail of citation distributions at intermediate times is only a transient shape.

Which feature of our model is responsible for *nonstationary* citation distributions? In what follows, we demonstrate that this is the probability of indirect citations P_0 which increases with the number of citations dependence according to Eq. 4.10. Indeed, we performed numerical simulation where we replaced the actual $P_0(K)$ dependence by a constant, $P_0 = 0.54$. Thus, our model reduces to a linear one, in which no parameter depends on K. Figure 8.5b shows that while the linear model accounts fairly well for early citation distributions, it fails miserably for late distributions. Indeed, the linear model yields citation distributions with the same slope for all times, while the slope of the measured citation distributions continuously decreases with time (Fig. 8.2).

8.3.2   Citation Lifetime

Previous empirical studies, which claimed the power-law citation distributions, implied that these distributions are scale-free [28, 41, 48]. Of course, since citations are discrete and non-negative, these distributions have a natural scale—the mean number of citations M. By a "scale-free citation distribution" one usually means the absence of the macroscopic scale besides the microscopic scale set by M. In particular, for Physics papers published in 1984, the mean number of citations accumulated by 2008 is $M = 26$. This scale is visible in Fig. 8.1c and it corresponds to transition from the convex part of the distribution for $K < M$ to the straight tail for $K > M$. This straight tail extends from $K/M \sim 3$ to $K/M = 230$. The huge disparity between these two numbers is considered as an indicator of the scale-free distribution. However, this is only an indicator and not a proof. In what follows we demonstrate that citation distributions do have a scale K_r associated with the onset of runaway papers. This hidden scale is barely visible in citation distributions but it pops out explicitly when we analyze citation trajectories of the papers.

To this end, we come back to our numerical simulation for 40,195 Physics papers published in 1984 and focus on citation trajectories of individual papers. By analyzing these trajectories, we found citation lifetime τ_0 in the same way as we did in Sect. 6 for the measured citation trajectories. Figure 8.6 shows the corresponding obsolescence rate, $\Gamma_{sim} = 1/\tau_0$. The simulated Γ_{sim} agrees fairly well with Γ_{meas}.

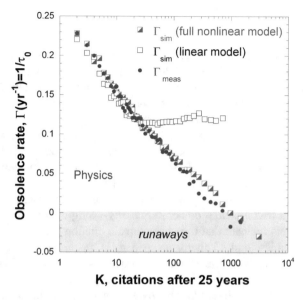

Fig. 8.6 The obsolescence rate of Physics papers, $\Gamma = 1/\tau_0$ versus K, the number of citations after 25 years. Blue circles show the results of our measurements from Fig. 6.6a. Red squares show results of the simulation based on full nonlinear model with the kernel given by Eq. 4.10. Γ_{meas} and Γ_{sim} are very close. Both decrease with increasing K and change sign at certain K_r. This decreasing trend is in agreement with our measurements. Open black squares show results of simulations based on a linear model with a constant kernel, $P_0 = 0.54$. Γ_{sim} decreases with K and achieves a constant level at $K \approx 20$. This simulation, which does not account for nonlinearity, disagrees with our measurements for $K > 20$. Reprinted with permission from Golosovsky [67]. Copyright (2017) by the American Physical Society. https://journals.aps.org/pre/abstract/10.1103/PhysRevE.96.032306

In particular, our model reproduces fairly well the decreasing $\Gamma(K)$ dependence and the divergence of citation lifetime at certain K_r. This K_r characterizes the onset of runaways and it can be discerned in the measured citation distribution as the point where the straight tail transforms to concave tail.

This surprising $\Gamma(K)$ dependence arises from the $P_0(K)$ dependence (Eq. 4.10). Indeed, numerical simulation using a linear model with $P_0 = const$ (Fig. 8.6b, open squares) shows that Γ_{sim} decreases with K only at small $K < 20$. This $\Gamma_{sim}(K)$ dependence arises from the fact that citations are discrete. Indeed, Eq. 5.1 describes a discrete Hawkes process where citation rate of a paper depends on the recent citation rate. This dependence introduces a positive feedback that amplifies fluctuations and it is responsible for decreasing $\Gamma_{sim}(K)$ dependence for small K. However, for $K > 20$ this purely statistical effect is washed out and $\Gamma_{sim}(K)$ dependence achieves a plateau. This is in contrast with the measured Γ_{meas} which continues to drop with increasing K for all K. Thus, while the linear model accounts for our measurements for $K < 20$, it disagrees with them for $K > 20$. Only nonlinear model fully accounts for the whole $\Gamma_{meas}(K)$ dependence.

8.4   Discussion

The nonstationary nature of citation distributions has been already noticed. Redner [144] found that citation lifetime of Physics papers increases with the number of citations. Lehmann et al. [94] analyzed the lifetime of high-energy physics papers, as defined through the fractions of the "live" and "dead" papers, and also found that the lifetime increases with the number of citations. Baumgartner and Leydesdorff [10] showed that citation trajectories of highly-cited papers are qualitatively different from the rest and do not come to saturation. The densification law established by Lescovec et al. [97] states that, as time passes, the growing citation networks do not rest self-similar but shrink in diameter and become denser.

In what follows, we find explanation of this surprising nonstationarity. First of all, there is a trivial source of nonstationarity—long citation lifetime. Indeed, our measurements indicate that the characteristic citation lifetime of the ordinary Physics, Economics, and Mathematics papers is $\tau_0 = 4.6, 9$ and 11.8 year, correspondingly (Fig. 6.6). The exponential process achieves saturation after $\sim 3\tau_0$. Thus, if we consider the time window of, say, 15 years, citation distributions for ordinary Physics papers should be already stationary, while those for Economics and Mathematics are not yet. Many citation distributions were measured in the time window of only 5–10 years after publication, when citation career of even the ordinary papers was not completed, hence it is not a surprise that such distributions were nonstationary and exhibited different shapes.

A more important source of nonstationarity is the fact that citation lifetime of a paper increases with the number of citations, and this holds for all studied research fields. When this number exceeds a certain critical number K_r, specific for each field and publication year, citation lifetime goes to infinity, in such a way that the papers with $K > K_r$ become runaways or supercritical papers [14]. We explain the origin of the runaway behavior within the framework of our copying/recursive search mechanism. Indeed, every newly published paper induces a train of citing papers whose authors find it in databases, Internet, etc. Each of these citing papers induces a cascade of secondary citing papers whose authors can copy the source paper into their reference list. Whether this cascade decays or propagates—this depends on the reproductive number $R_{rep} = P_0 N^{nn}$, where N^{nn} is the average number of the second-generation citing papers per one first-generation citing paper (the fan-out coefficient), and P_0 is the probability of indirect citation (copying). For ordinary papers, $R_{rep} < 1$ and the citation cascade decays, while for highly-cited papers, $R_{rep} > 1$ and the citation cascade propagates indefinitely in time. The papers that managed to overcome the tipping point of $R_{rep} = 1$ are runaways and their presence can result in a "winner-takes-all" network [88, 172].

What are the implications of nonstationarity of citation network with respect to its degree distribution? If we focus on the set of papers in one field published in one year, its citation distribution at early times mimics the fitness distribution, while at later times it develops according to our model (Eq. 5.1). While initial citation distribution is close to log-normal, the evolving distributions straighten and become closer to power-law. This corroborates the conjecture of Mokryn and Reznik [117] who assumed that the power-law degree distribution can be found only in static networks, namely, those that underwent a long period of development. We show here that citation distribution in the long time limit, once considered as static, is, in fact, always transient. As time passes, its tail evolves from the convex shape to straight line, and then it can even acquire the concave shape. The rate of this evolution is field-specific and depends on fitness distribution, on the growth of the number of publications in this field, and on the average reference list length.

So, what is the source of the "power-law" citation distributions? Our findings indicate that these are inherited from the fitness distribution and then modified during citation process. The question of why citation distributions follow the power-law dependence is thus relegated to fitness distribution. The deep question is why fitness distribution has certain shape and this shape seems to be universal? This is an open question and we touch it briefly in the end of Chap. 6.

Does the power-law degree distribution necessarily imply a scale-free network? The notion of the scale-free networks was introduced by Barabasi and Albert [3] who realized that the ubiquity of the power-law degree distributions in complex networks implies some universal generating mechanism. The Barabasi-Albert preferential attachment mechanism generates a growing complex network whose degree distribution becomes stationary in the long time limit. This stationary shape is the power-law which implies the scale-free network. Before the advent of complex networks, the scale-free phenomena appeared in the condensed matter physics and have been usually associated with phase transitions and critical points. In condensed matter physics, the characteristic scale associated with this phenomenon— correlation length—diverges at critical point and this results in universal power laws for static and dynamic properties of the substance. The universal power-law degree distribution in complex networks has a great appeal to physicists with their quest for grand unified theories, since it implies that the properties of such diverse objects—complex networks and substances at critical point—are described by the same mathematical formalism. However, the equivalence between the power-law degree distribution and the scale-free character of a complex network holds only if the degree distribution is stationary or at least does not change its shape. However, for nonstationary networks there can be a dynamic scale that governs the network development. We demonstrate here that citation distributions are nonstationary, their shape changes with time, and there is a certain dynamic scale that is important for network growth. Therefore, a complex network may be non-scale-free even though its degree distribution follows a power-law dependence.

Can our findings be extended to other complex networks? We found that nonstationary citation distribution is related to nonlinear citation dynamics. Citation dynamics of patents are also nonlinear [44, 76, 186]. Thus, we expect that patent citation distribution is nonstationary as well. With respect to degree distribution in Internet networks—the latter have an important distinction: unlike citation networks which are directed and acyclic, the WWW is not and the links there can be edited. Nevertheless, runaways were detected in the distribution of the Web page popularity as well [87].

In summary, we have demonstrated that while statistical distribution of citations to scientific papers can be fitted by the power-law dependence, this distribution is nonstationary and does not acquire a limiting shape in the long time limit. While the power-law fit implies the scale-free citation network, we show that there is a hidden dynamic scale associated with the onset of runaways. Thus, the similarity between the measured citation distributions and the power-law dependence does not have deep implications.

Chapter 9
Comparison to Existing Models

Abstract We make a survey of models of citation dynamics and focus on the preferential attachment and fitness models. We show that under certain realistic conditions these models are equivalent. In order to find the microscopic foundations of the preferential attachment mechanism, we analyze theoretically and experimentally several citation networks and demonstrate that, for a broad fitness distribution, this mechanism reduces to the fitness model. The fitness model yields the long-sought explanation for the initial attractivity K_0, an elusive parameter which was left unexplained within the framework of the empirical preferential attachment model. We show that the initial attractivity is determined by the width of the fitness distribution. We compare the preferential attachment and fitness models to our microscopic model of citation dynamics based on recursive search and show that our model contains both these phenomenological models.

Keywords Citation models · Preferential attachment · Fitness model · Recursive search

9.1 Preferential Attachment Mechanism

9.1.1 Theoretical Model

To explain the power-law distribution of citations of scientific papers, de Solla Price suggested his cumulative advantage model [138]. This model assumes a network consisting of papers that appear with constant rate N, each paper extending $\sim R_0$ references to older ones. The probability of a new paper i to cite an old paper j is

$$\Pi_{ij} \propto (K_j + K_0), \tag{9.1}$$

where K_j is the number of citations of paper j. The initial attractivity K_0 ensures that new papers become cited as well (de Solla Price denoted K_0 by c). The cumulative advantage mechanism captured by Eq. 9.1 yields a power-law citation distribution, $p(K) \sim K^{-\nu}$, with the exponent

© The Author(s), under exclusive license to Springer Nature Switzerland AG 2019   93
M. Golosovsky, *Citation Analysis and Dynamics of Citation Networks*,
SpringerBriefs in Complexity, https://doi.org/10.1007/978-3-030-28169-4_9

$$\nu = 2 + \frac{K_0}{R_0}. \tag{9.2}$$

In the absence of any clue, Price postulated $K_0 = 1$. Since $R_0 \gg 1$, Eq. 9.2 yields ν slightly exceeding 2.

Following proliferation of digitized information in 1990s, a number of information, biological, and social complex networks came to the forefront of scientific research, most of them exhibiting power-law degree distributions with $\nu \sim 3$ [7, 15, 28, 121]. To account for these distributions, Barabasi and Albert suggested the preferential attachment model [3] which is very similar to but not identical with the Price's cumulative advantage. With regards to citation network, the Barabasi-Albert model assumes that a new paper i cites an old paper j with probability

$$\Pi_{ij} \sim (K_j + R_0). \tag{9.3}$$

Equation 9.3 yields $\nu = 3$ and this value better fits the observations.

In what follows, we do not make distinction between the Price's and Barabasi-Albert approaches and consider Eq. 9.1 with unspecified K_0 as the preferential attachment model. Its success in explaining the seemingly universal power-law degree distribution in complex networks prompted a flurry of theoretical generalizations, including nonlinear attachment rule [88] and aging [47, 73, 127, 176, 178]. The generalized preferential attachment model is captured by the following equation,

$$\Pi_{ij} \propto A(t)(K_j + K_0)^\zeta, \tag{9.4}$$

where $t = t_i - t_j$ is the age of paper j with respect to paper i, ζ is the attachment exponent, and $A(t)$ is the aging function which is different from that appearing in the model of Chap. 3.

9.1.2 Model Validation by Measurements

Straightforward verification of the preferential attachment model, as given by Eq. 9.4, requires analysis of decisions made by the authors of the papers and, to the best of our knowledge, this approach has not been implemented so far. The common approach is to trace citation dynamics of individual papers. To this end, the perspective is shifted from the citing paper to the cited paper and this results in the following equation,

$$\Delta K_j = \tilde{A}(t_j)(K_j + K_0)^\zeta \Delta t, \tag{9.5}$$

where $\Delta K_j = k_j(t)\Delta t$, $k_j(t)$ is the citation rate of the paper j, the aging function is $\tilde{A}(t) = \frac{A(t)}{\sum_l A(t_l)[K_l+K_0]^\zeta} R_0 N$, and N is the annual number of publications. [Note difference between $A(t)$ and $\tilde{A}(t)$: for the Barabasi-Albert model $A(t) = 1$ while $\tilde{A}(t) = \frac{2N}{t}$. Both $A(t)$ and $\tilde{A}(t)$ are different from the aging functions appearing in the model of Chap. 3.]

The measurements of Refs. [90, 95, 113, 131, 134] verified that the growth of citation networks indeed follows Eq. 9.5 although some of these measurements exhibited preferential attachment (ΔK_j grows with K_j) only for papers with low and moderate K_j while the highly-cited papers frequently exhibit anti-preferential attachment (ΔK_j decreases with K_j) [32, 90].

With respect to linearity, early measurements claimed linear or close-to-linear ($\zeta \approx 1$) preferential attachment [83, 122] while subsequent measurements for large datasets of scientific papers [70, 77] and patent citations [44, 76] revealed *superlinear attachment* with the exponent $\zeta \sim 1.25$.

The measurements of *initial attractivity* posed a significant challenge to the preferential attachment model. Indeed, for citations to scientific papers, the initial attractivity is very small $K_0 \sim 1$ [71, 76]. Patent citations also yield small $K_0 \sim 1$ [44, 77]. Such small initial attractivioty better conforms to Price's conjecture $K_0 = 1$ than to the Barabasi-Albert conjecture, $K_0 = R_0$. However, $K_0 \sim 1$ yields too small an exponent of the power-law citation distribution which is incompatible with observations [44, 49, 71, 83].

Several specific predictions of the preferential attachment model are also at odds with observations. In particular, this model predicts *the first mover advantage*, namely, strong positive correlation between the paper's age and the number of citations it garners. However, the measurements reveal only weak such correlation [124, 125]. The model predicts that citation dynamics of papers decelerates with time and the *citation trajectories* of the papers published in the same year should be very similar. However, the measurements show that these trajectories strongly diverge [70, 87] and do not necessarily decelerate with time. In particular, there are "sleeping beauties" [84], the papers whose citation trajectories accelerate with time. Also, the preferential attachment model predicts that *citation distributions for the papers published in one year* should be narrow and close to the exponential [7, 121], while the measurements show that these distributions are wide and close to the power-law or log-normal [72, 87, 144]. Finally, the preferential attachment model shows that the *assortativity* of the growing complex network is determined by the initial attractivity K_0, in such a way that for small and positive initial attractivity (as it has been measured in many networks) the network should be disassortative [7, 122]. This does not fit measurements which reveal that citation networks are weakly assortative [7, 60, 72, 149, 180].

9.2 Fitness-Based Preferential Attachment

When the preferential attachment model is applied for quantitative account of real growing complex networks, it meets too many difficulties. The most popular solution is to introduce fitness [14], an attribute that characterizes the propensity of a node to attract edges (the ability of a paper to draw citations). The fitness is an empirical attribute that should be found from measurements, although there were several attempts to associate it with node relevance [107], local clustering coefficient [5], node's rank [54], and PageRank coefficient [187].

9.2.1 Multiplicative Fitness

How fitness can be incorporated into dynamic equation of network growth? The Bianconi-Barabasi model [14] introduces fitness *on top of the preferential attachment*. In the context of citation networks, this model postulates that the probability of paper i to cite paper j is the product of paper's fitness η_j and the number of its previous citations, namely

$$\Pi_{ij} \propto \eta_j K_j. \tag{9.6}$$

Solution of Eq. 9.6 yields citation trajectories $K_j(t)$ which strongly depend on fitness, in such a way that a high-fitness latecomer can outperform a low-fitness old paper. Thus, fitness solves the problem of the first-mover advantage. It solves other problems as well, in particular, citation distribution for the papers of the same age is determined more by fitness distribution rather than by the number of previous citations. The most striking prediction of the Bianconi-Barabasi model is the presence of supercritical papers that take a lion share of all citations. Such supercritical papers were indeed observed [9, 67] and this non-intuitive prediction brought a wide popularity to the Bianconi-Barabasi model [12, 28, 33, 61, 175].

However, Eq. 9.6 represents only conceptual model. To convert it into a quantitative tool that accounts for real-life citation dynamics, Wang et al. [175] added initial attractivity K_0 and aging constant $A(t)$, while Pham et al. [134, 135] added nonlinearity. The most general form of Eq. 9.6 reads

$$\Pi_{ij} = \eta_j A_j(t)(K_j + K_0)^\xi. \tag{9.7}$$

Wang, Song, and Barabasi parametrize their aging function by the log-normal representation [they denoted it by $P_j(t)$ while we denote it here by $A_j(t)$], $A_j(t) = \frac{1}{\sqrt{2\pi}\sigma_j t} e^{-\left(\frac{(\ln t - \mu_j)^2}{2\sigma_j^2}\right)}$ where μ_j and σ_j are parameters which are specific for each paper.

While Eq. 9.7 solved many problems, its success came at the price of too many empirical parameters which are required to describe citation trajectory of a paper: in addition to the number of previous citations K_j and age t_j, Eq. 9.7 adds fitness η_j, exponent ζ, and two parameters μ_j, and σ_j which characterize the aging function for each paper.

9.2.2 Additive Fitness

The fitness can be also introduced through optimization procedure (Ref. [128]) or through the following equation [11, 49, 50, 108]

$$\Pi_{ij} \propto (K_j + \eta_j), \tag{9.8}$$

which is nothing else but the preferential attachment model in which fitness η_j replaces the initial attractivity K_0.

The growth dynamics described by Eqs. 9.7 and 9.8 are not that different as it could seem. In fact, the combination of nonlinear preferential attachment (Eq. 9.5) with additive fitness (Eq. 9.8) mimics Eq. 9.6, in particular, it yields supercritical nodes. To demonstrate this, we adopt continuous approximation and replace ΔK_j in Eq. 9.5 by $\frac{dK_j}{dt}\Delta t$. In view of Eq. 9.8, Eq. 9.5 can be recast as follows:

$$\frac{dK}{dt} = \tilde{A}(t)(K + \eta)^\zeta, \tag{9.9}$$

where we replaced K_0 by η and dropped index j, for brevity. We introduce $\varepsilon = \zeta - 1$, solve Eq. 9.9 for $\varepsilon > 0$, and find

$$K(t) = \frac{\eta}{\left[1 - \epsilon\eta^\epsilon \int_0^t \tilde{A}(\tau)d\tau\right]^{\frac{1}{\epsilon}}} - \eta. \tag{9.10}$$

To analyze Eq. 9.10, we assume that the integral $\int_0^t \tilde{A}(\tau)d\tau$ converges as $t \to \infty$. This assumption allows us to introduce $\eta_{crit} = \left[\epsilon \int_0^\infty \tilde{A}(\tau)d\tau\right]^{-\frac{1}{\epsilon}}$, in such a way that Eq. 9.10 reduces to

$$K(t) = \frac{\eta}{\left[1 - \left(\frac{\eta}{\eta_{crit}}\right)^\epsilon \frac{\int_0^t \tilde{A}(\tau)d\tau}{\int_0^\infty \tilde{A}(\tau)d\tau}\right]^{\frac{1}{\epsilon}}} - \eta, \tag{9.11}$$

where t is the paper's age. For $\eta < \eta_{crit}$, Eq. 9.11 yields $K(t)$ that increases with time and eventually achieves saturation, $K(\infty) = \frac{\eta}{\left[1 - \left(\frac{\eta}{\eta_{crit}}\right)^\epsilon\right]^{\frac{1}{\epsilon}}} - \eta$. However, for $\eta \geq \eta_{crit}$, $K(t)$ does not achieve saturation and the node becomes supercritical, namely, the number of its citations undergoes a finite-time singularity at certain t_0, in such a way that Eqs. 9.10 and 9.11 hold only for $t < t_0$. Thus, for superlinear preferential attachment, $\epsilon > 0$, Eq. 9.8 predicts the supercritical nodes—exactly as Eq. 9.6 does.

9.3 Fitness-Only Models

The fitness model, suggested by Caldarelli et al. [29] and further developed in Refs. [11, 61, 104, 126, 154], considers a general complex network. This model departs from the idea of preferential attachment and assumes that the probability of attachment between a new node i and the target node j depends only on their fitnesses, η_i and η_j, and does not depend on their degrees. For citation networks, Ref. [29] assumed that the probability of paper i to cite paper j is $\Pi_{ij} = f(\eta_i, \eta_j)$ where η_i and η_j are paper's fitnesses, and $f(\eta_i, \eta_j)$ is the symmetric function of its arguments (linking function). Ref. [29] considered additive linking function but the later publication of the same group [150] introduced multiplicative linking function, $f(\eta_i, \eta_j) \sim \eta_i \eta_j$. The latter assumption became more popular and it allows the following generalization. Consider an old paper j. If fitness is determined by similarity and all papers under consideration belong to the same field, then the paper j will garner citations with the rate $\Delta K_j \propto \overline{\eta_i} \eta_j$ where $\overline{\eta_i}$ is the average fitness of new papers. This average fitness can be absorbed into the aging function, in such a way that the probability of a new paper i to cite an older paper j is

$$\Pi_{ij} \sim \eta_j A(t_j). \tag{9.12}$$

What is fitness? On the one hand, the fitness includes the notion of similarity known as homophily in social networks [20]. Indeed, citation networks consist of communities and subcommunities. New papers tend to attach to similar papers, those belonging to the same community. To measure similarity, Refs. [39, 108] suggested to use overlap of contents or bibliographies, while Refs. [101, 113, 122] suggested to use the overlap of common neighbors. Another ingredient of fitness is associated with quality or talent. This component is not easy to estimate when the paper first appears, it can be measured only after it already garnered some citations.

9.4 Explanatory Models

The preferential attachment model is phenomenological, namely, it makes plausible but unsubstantiated assumptions regarding the mechanism by which the papers acquire citations, in particular, it deduces this mechanism (rich-gets-richer) from the citation distributions. According to the preferential attachment model, the algorithm used by an author of a new paper to choose his references is as follows. The author considers the number of citations of all papers and chooses a paper to cite accordingly. Thus, each author shall know the number of citations garnered by all papers. However, in the pre-Internet era this number was known only to those few authors who had access to Science Citation Index. [After appearance of Internet and especially Google Scholar, the number of citations can be found using a few clicks. Once the preferential attachment model gained popularity in the scientific community, it became a self-fulfilling prophecy but its microscopic origin in the pre-internet era remains obscure, at least for citation networks.]

Moreover, the preferential attachment mechanism of network growth presents a major conceptual difficulty because it is global and not local. Indeed, Eqs. 9.5 and 9.7 imply that each incoming node shall know degrees of all other nodes. Although this can be true for collaboration and some other social networks [35, 36], in general, a new node has bounded knowledge—it is familiar only with a limited set of nodes. With the proliferation of informational databases such as Google Scholar, Scopus, ISI Web of Science, etc., global information on many complex networks became easily accessible. In particular, for citation networks, the incentive to cite a certain paper nowadays may indeed come from the number of previous citations. Thus, the preferential attachment model is becoming a self-fulfilling prophecy.

While we showed here that the fitness model explains the initial attractivity, it can hardly serve as a basis for the explanation of citation network growth since it is too phenomenological and devoid of specific details characterizing real networks.

There are several non-phenomenological models of citation dynamics based on a some realistic algorithm which the authors use to choose their references. The most important class of such models is the recursive search [154, 170] also known as link copying or redirection [89], random walk or local search [66, 82], triple (triangle) formation [179], triadic closure [106], or forest fire model [96, 162]. This algorithm assumes that an author of a new paper chooses randomly an older paper and includes it into his reference list. Then he explores the reference list of the newly chosen paper and copies one [82] or all [89, 92] of its references. This is a one-level recursive search while Refs. [91, 96, 162] considered a multilevel recursive search, whereby an author explores reference lists of all previously chosen papers. Vazquez [170] showed that one-level recursive search mechanism reduces to the following probability of a new paper i to cite a target paper j,

$$\Pi_{ij} = \lambda + q K_j. \tag{9.13}$$

Here, λ is the probability of random search, $q K_j$ is the probability of recursive search, and K_j is the number of citations of the target paper. Thus, Eq. 9.13 reduces to Eq. 9.1 and it was long considered as a microscopic justification of the preferential attachment mechanism. Note, however, that the derivation of Eq. 9.13 by Ref. [170] was based on very simplifying assumptions: all papers have the same probability of being randomly-chosen, only one ancestor of the randomly-found paper is chosen by a new paper, there is no aging and no memory, etc. Of these assumptions, the absence of aging is crucial for the derivation of Eq. 9.13. In the presence of aging, the recursive search mechanism yields

$$\Pi_{ij} = \lambda + q \Delta K_j(t - \Delta) \tag{9.14}$$

where $\Delta K_j(t - \Delta)$ is the number of citations garnered by the paper in the time window $t, t - \Delta$ and Δ is the characteristic memory span. As we have shown in previous chapters, Eq. 9.14 captures citation dynamic of papers more realistically than Eq. 9.13. However, this equation is very different from the preferential attachment model, especially for short Δ.

In this book we have demonstrated a realistic model of citation dynamics which embeds fitness-based search into recursive search and takes proper account of the aging, memory, and topology of citation network. In fact, we put more flesh on the bones of the recursive search algorithm developed by Vasquez [170] and converted it into quantitative tool to account for citation dynamics of scientific papers [72]. Our model combines the fitness model and the preferential attachment model (albeit with memory, following approach of Refs. [65, 109, 118, 127, 145, 176]). This model assumes a reasonable algorithm which the author follows when composes his reference list. Although our model does not assume preferential attachment, citation dynamics generated by this model follows Eq. 9.22 as all fitness-based models do. Comparison of Figs. 5.5 and 9.2 shows this quite convincingly. These two figures look very similar although the latter was generated using a simple fitness model and the former was generated using our recursive search model. And both of them look as if they were generated using a preferential attachment model!

9.5 Equivalence Between Preferential Attachment and Fitness Models

While both Caldarelli and Bianconi-Barabasi fitness models capture the node's quality/similarity, they are mathematically different. Indeed, the Bianconi-Barabasi's fitness has been introduced on top of the preferential attachment through Eq. 9.6, while Caldarelli's fitness has nothing to do with the preferential attachment. We claim, however, that the Caldarelli's fitness-only model is in some sense equivalent to the classical preferential attachment model.

To demonstrate this equivalence, we consider a complex network that grows according to Eq. 9.12. Every node is endowed with a certain fitness η that remains constant during node's life. This fitness is drawn from the fitness distribution $\rho(\eta)$ where $\int_0^\infty \rho(\eta)d\eta = 1$. We also assume that ΔK, the number of new edges garnered by a node during time window $(t, t + \Delta t)$, is represented by the Poisson distribution, $\frac{\lambda^{\Delta K}}{\Delta K!}e^{-\lambda}$, where Poissonian rate,

$$\lambda = \eta A(t)\Delta t, \tag{9.15}$$

is determined by the node's fitness η. $A(t)$ is the normalized aging function. Under this constraint, the fitness η is the node's degree in the long-time limit, namely, $\eta \approx K^\infty$.

Since Eq. 9.15 is memoryless, the number of edges that each node garners through the period from $t = 0$ to t also follows Poisson distribution with the node-specific rate

$$\Lambda = \eta \int_0^t A(\tau)d\tau. \tag{9.16}$$

We consider the set of N nodes that joined the network at the same moment which we set as $t = 0$. Among these, we focus on the subset of nodes that garnered K edges by time t. Their number is

$$n(K, t) = N \int_0^\infty \frac{\Lambda^K}{K!} e^{-\Lambda} \rho(\eta)d\eta, \tag{9.17}$$

where Λ depends on η in accordance with Eq. 9.16. During time window $(t, t + \Delta t)$, each of these $n(K, t)$ nodes garners $\sim \lambda$ edges, in such a way that the average number of new edges garnered by a node from this subset is

$$\overline{\Delta K} = \frac{N \int_0^\infty \lambda \frac{\Lambda^K}{K!} e^{-\Lambda} \rho(\eta)d\eta}{n(K, t)}. \tag{9.18}$$

We substitute Eq. 9.15 into Eq. 9.18, note that $\lambda = \Lambda \tilde{A}(t)$, where $\tilde{A}(t) = \frac{A(t)}{\int_0^t A(\tau)d\tau}$, use the equality

$$\Lambda Poiss(\Lambda, K) = (K + 1)Poiss(\Lambda, K + 1), \tag{9.19}$$

and come to

$$\overline{\Delta K} = \tilde{A}(t)(K + 1)\frac{n(K + 1, t)}{n(K, t)}\Delta t. \tag{9.20}$$

Here, $n(K + 1, t)$ is the number of nodes that garnered $K + 1$ edges by time t. For broad fitness distribution and for $K \gg 1$, $n(K + 1, t) \approx n(K, t)$, in such a way that Eq. 9.20 reduces to

$$\overline{\Delta K} = \tilde{A}(t)(K + 1)\Delta t. \tag{9.21}$$

This expression is nothing else but Eq. 9.4 with $K_0 = 1$ and $\epsilon = 0$. [Interestingly, this derivation validates the initial conjecture de Solla Price, $K_0 = 1$.] A similar result was obtained earlier by Burrell [25] using a different approach. Equation 9.21 is commonly accepted as an evaluation tool of growth mechanism of real complex networks. We demonstrate here that this equation is undiscriminating and holds not only for the preferential attachment but for the fitness model as well.

To validate Eqs. 9.20 and 9.21 through numerical simulation, we considered a set of 400,000 nodes with a log-normal fitness distribution $\rho(\eta) = \frac{1}{\sqrt{2\pi}\sigma\eta} e^{-\frac{(\ln\eta - \mu)^2}{2\sigma^2}}$ where $\mu = 1.63$ and $\sigma = 1.12$ (We defined here fitness differently from what was defined in Chaps. 3 and 7. Here, η is multiplied by R_0, the average reference list length). We simulated the growth of these nodes using Eq. 9.15 and the aging function $A(t) = \frac{0.035t}{|t-2.4|^{1.3}}$. (The parameters of the simulation were chosen in such a way as to imitate measured citation dynamics of Physics papers as reported in Ref. [72].) The time was run from $t = 0$ to $t = 25$ with steps $\Delta t = 1$, in such a way that $\sum_0^{t=25} A(t) = 1$. For each node j in this set, we determined $K_j(t)$, the total number of edges accumulated after time t, and $\Delta K_j(t)$, the number of additional edges gained between t and $t + 1$. For every t, we grouped all nodes into 40 logarithmically-spaced bins, each bin containing the nodes with close values of K. For each bin, we determined ΔK-distribution and found its mean, $\overline{\Delta K}$. Figure 9.1a plots $\overline{\Delta K}$ versus K for small K. We observe straight lines with common intercept of -1, as suggested by Eq. 9.21. To fit the whole $\overline{\Delta K}(K)$ dependence, we used the following equation

$$\overline{\Delta K} = \tilde{A}(t)(K + K_0), \tag{9.22}$$

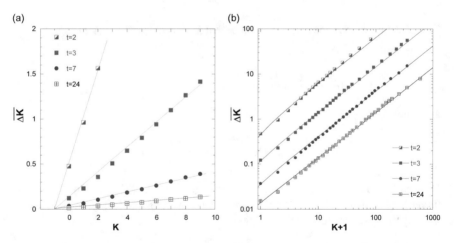

Fig. 9.1 Numerical simulation of the growth dynamics of 400,000 nodes with the aging function $A(t) = \frac{0.035t}{|t-2.4|^{1.3}}$ and log-normal fitness distribution with $\mu = 1.63$ and $\sigma = 1.12$. ΔK is the mean growth rate, K is the number of accumulated edges, and t is the age. (**a**) $\overline{\Delta K}$ versus K for small K. The data for all t lay on straight lines with common intercept of -1, suggesting $K_0 \approx 1$. (**b**) $\overline{\Delta K}$ versus K for all K. Continuous lines show fits to Eq. 9.21 with $K_0 = 0.7, 0.8, 0.85$ and 1 for $t = 2, 3, 7$ and 24, correspondingly. Reprinted with permission from Golosovsky [68]. Copyright (2018) by the American Physical Society. https://journals.aps.org/pre/pdf/10.1103/PhysRevE.97. 062310

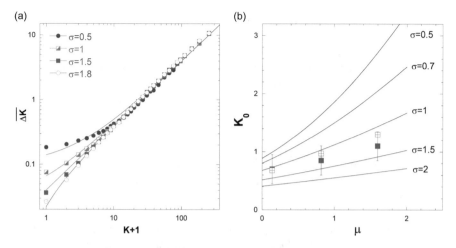

Fig. 9.2 (a) Numerical simulation of the growth dynamics of the set of 400,000 nodes with different log-normal fitness distributions having the same $\mu = 1.63$ and different σ. The symbols show results of numerical simulation, continuous lines show linear approximation $\overline{\Delta K} = A(K + K_0)$ with $A = 0.04$ and $K_0 = 3.5, 1.65, 1,$ and 0.55 for $\sigma = 0.5, 1, 1.5,$ and 1.8, correspondingly. (b) Initial attractivity K_0 estimated from Eq. 9.24 in dependence of the parameters of the log-normal fitness distribution, μ and σ. The filled squares show our measurements for Physics, Economics, and Mathematics papers published in 1984 (see Ref. [70]), the open circles show our expectations based on measured μ and σ of the log-normal fitness distribution for these very datasets. The measured values of K_0 are close to those predicted by Eq. 9.24. Reprinted with permission from Ref [68]. Copyright (2018) by the American Physical Society. https://journals.aps.org/pre/pdf/10.1103/PhysRevE.97.062310

where K_0 is the fitting parameter. Figure 9.1b shows that this equation fits the data fairly well for all K.

Figure 9.2 shows $\overline{\Delta K}$ versus $(K + K_0)$ dependences for log-normal fitness distributions with different σ and for time slices $\Delta t = 1$. These dependences are also well fitted by Eq. 9.22 with $K_0 \sim 1$.

To estimate K_0 from the data in a more precise way, we turn to Eq. 9.22. It indicates that at small K, $\overline{\Delta K} \rightarrow \tilde{A}(t)K_0$. On another hand, Eq. 9.20 yields for $\Delta t = 1$

$$\overline{\Delta K}|_{K=0} = \tilde{A}(t)\frac{n(1, t)}{n(0, t)}. \tag{9.23}$$

Thus,

$$K_0 \approx \frac{n(1, t)}{n(0, t)} = \frac{\int_0^\infty \Lambda e^{-\Lambda}\rho(\Lambda)d\Lambda}{\int_0^\infty e^{-\Lambda}\rho(\Lambda)d\Lambda}. \tag{9.24}$$

where Λ is given by Eq. 9.16. We note that $\rho(\Lambda)$ follows the log-normal distribution which is nothing else but the fitness distribution with shifted mean, $\mu' = \mu +$

$\ln(\int_0^t A(\tau)d\tau)$. Since $\int_0^t A(\tau)d\tau \to 1$ in the long time limit, the difference between μ and μ' becomes increasingly smaller at long t. Figure 9.2b shows K_0 calculated according to Eq. 9.24 as a function of μ and σ. We observe that K_0 increases with μ and decreases with σ. These dependences can be captured by the approximate empirical expression

$$K_0 \approx \frac{e^{\frac{\mu}{1+\sigma}}}{(1+\sigma^2)^{0.6}}. \tag{9.25}$$

For reasonable values of μ from 0 to 2 and σ from 1 to 2, K_0 lies between 0.5 and 1.5. It is determined by σ, and, to a lesser extent, by μ. All this means the following: if K_0 is measured using Eq. 9.22 using extrapolation from large K, one always gets $K_0 = 1$. On another hand, since most fitness distributions are broad, then the estimates made using Eq. 9.22 for small K, as it is usually done in most studies, yield $K_0 = 0.5$–1.5. Figure 9.2b shows that for narrow fitness distribution, K_0 can be higher.

We plot on Fig. 9.2b the measured values of K_0 which were inferred from our studies of citation dynamics. We considered three research fields: Physics, Economics, and Mathematics and found that the fitness distributions for all these fields are log-normals with the same $\sigma = 1.1$. The measured and calculated initial attractivities K_0 are in good agreement and are all close to 1.

Thus, our numerical simulation supports Eq. 9.22 with initial attractivity $K_0 \approx 1$, as it was postulated by de Solla Price [138]. The natural question arises—why $K_0 \approx 1$ is so widespread? Figure 9.2b shows that $K_0 \approx 1$ corresponds to $\sigma = 1$–1.5 irrespective of μ. Nguyen and Tran [126] used numerical simulation to study complex networks with log-normal fitness distribution that grow according to Eq. 9.12. They found that the resulting network structure strongly depends on the width of the fitness distribution σ, in particular, the power-law degree distribution appears only for $\sigma \approx 1$ and its exponent ν is close to 3. This observation implies that the initial attractivity is coupled to the exponent of the degree distribution, in such a way that within the framework of the fitness model, the universality of $K_0 \approx 1$ in complex networks is a consequence of the fact that most of them exhibit power-law degree distributions with $\nu \sim 3$.

9.6 The Genuine Preferential Attachment Exists and is Related to Nonlinear Citation Dynamics

Although our model is based on the measurements on citation networks, we believe that it is more general and relevant to other complex networks as well. In what follows, we present our reflections on this subject. Our analysis shows that if network growth is considered from the perspective of a target node and is studied using the mean-field approximation, namely, by averaging over many similar nodes, one cannot distinguish between the preferential attachment and the fitness models—both of them yield Eq. 9.22. Thus, in all that concerns the mean-field network dynamics, preferential attachment is equivalent to fitness model, in other words, the rich-gets-richer mechanism reduces to the fit-gets-richer (good-gets-richer) mechanism [29, 135]. This is surprising since these two models are based on different premises. The preferential attachment model assumes that all nodes are born equal, the inequality in their degree coming by chance. After this inequality has been established, it is amplified by the autocatalytic process represented by Eq. 9.1. In contrast, the fitness model and the fitness-based recursive search model assume that the nodes are born unequal, each newly born node is endowed with a certain fitness. The latent fitness inequality becomes evident when the nodes have been developing for some time. Surprisingly, the two opposing assumptions underlying network growth—all nodes are born equal or different—result in the same Eq. 9.22.

This does not mean that the two models are equivalent. While the preferential attachment model does not specify the initial attractivity, the fitness model with aging explains it perfectly well—it is determined by the shape of the fitness distribution. With respect to the power-law degree distribution in complex networks: the preferential attachment relates its to the strategy by which the new node attaches to old nodes, while the fitness model implies that this distribution is inherited from the fitness distribution. The fitness model successfully explains the first-mover advantage, degree distribution for the nodes of the same age, different trajectories of the nodes of the same age, etc. However, this model does not account for the nonlinear growth commonly observed in citation networks.

Although it could seem that the fitness model is a more appropriate framework to conceptualize network growth, Eqs. 9.1 and 9.22 can still be valid since the preferential attachment is a structural rather than explanatory model. Indeed, the relation $\Pi_{ij} \sim K_j$ does not imply that a new node i crawls through the whole network in order to gain information about degrees of all other nodes j. What occurs in reality is that the network grows following some local rule and this rule becomes imprinted in the network topology. When the network growth is analyzed, the changes in topology are visible while the underlying microscopic growth rule is not. This feeds the illusion that the growth dynamics is determined by network topology while in reality the reverse is true.

The challenge is to uncover the microscopic rules of network growth that produce the given network topology. We showed that the recursive search is one of the plausible microscopic mechanisms of network growth. What is the relation of this mechanism to the genuine preferential attachment—namely, the algorithm whereby a new node finds well-connected older nodes and attaches to them? It has been generally believed that the recursive search is one of realizations of this algorithm, since if a new node makes a random choice among the neighbors of already chosen nodes, it has high probability of picking up highly-connected nodes. We demonstrate here (Eq. 3.24) that this strategy works in a straightforward way only if the recursive search does not have memory. In reality, recursive search has rather short memory [72], and it is not clear whether highly-connected nodes can be found by this simple strategy: random choice among the neighbors of already chosen nodes. In Chap. 4

we showed that the recursive search there follows a more clever strategy: the search in the network neighborhood of the previously chosen nodes is not random but has preference for those neighbors that are connected to several already chosen nodes. The cartoon picture of this strategy is as follows. Simple recursive search: if Alice is linked to Bob, and Bob is linked to Frank, there is a chance that Alice will link to Frank. Clever recursive search: if Alice is linked to Bob and Charlie, and both of them are linked to Frank, then Alice will link to Frank almost for sure. Thus, if a new node identifies a target node in the network vicinity of the two or more previously chosen nodes, the probability of attachment to such node exceeds the sum of probabilities for each path, namely, multiple paths interfere constructively, reinforcing one another. The synergetic interaction between the paths to the next-nearest neighbors ensures that a new node finds highly-connected nodes. This strategy of exploring next-nearest neighbors can still be considered as a local strategy, but in fact, it is one step towards global search and this is one of the ways how the genuine preferential attachment emerges within the framework of the recursive search model.

Appendix A
Details of Numerical Simulation

In our numerical simulations of citation dynamics of 40,195 Physics papers published in 1984 we used the following parameters. $\gamma + \beta = 1.2 \text{ yr}^{-1}$, as found in our measurements of indirect references and citations; $m_{dir}(t)$ from Fig. 10 (see main text). We assumed that $n^{nn}(t)$ dependence mimics $m(t)$ dependence, namely $n^{nn}(t) = \frac{m(t)}{\bar{s}}$ where $\bar{s} = 1.2$ is the average over all Physics papers published in 1984 and $M(t)$ is shown in Fig. 6 (main text). With respect to $\tilde{P}_0(K)$, it is given by the following expression: $\tilde{P}_0(K) = Pf(K)$ where $P = 0.34$ and $f(K) = 1 + 0.82 \log K$. Here, $P = 0.34$ characterizes the probability of indirect citations for low-cited papers, and $f(K)$ stays for logarithmic correction which is most important for highly-cited papers.

We can assemble all time-dependent functions in the kernel together, in such a way that Eq. 4.9 reduces to

$$\lambda_i(t) = \eta_i m_{dir}(t) + \sum_{\tau=1}^{t} f(K_i) F(t - \tau) k_i(\tau) \tag{A.1}$$

where $F(t) = Pn^{nn}(t)e^{-(\gamma+\beta)t}$. For the Physics papers published in 1984 we find $F(1) = 0.089$, $F(2) = 0.138$, $F(3) = 0.046$, $F(4) = 0.012$, $F(5) = 0.0035...$ Thus, at first F grows with time as the paper receives more recognition (there is approximately one year delay between the publication of the paper and its first citation) and then decays exponentially. This obsolescence is strong, hence the memory of the citation process is restricted to a few years.

In our numerical simulations we used Eq. A.1. For approximate calculations one can also use a simplified numerical scheme according to which Eq. A.1 is considered as an autoregressive model. We looked for the model of minimum order that can faithfully represent our measurements and found that the first-order model is unsatisfactory while the second-order autoregressive model

$$\lambda_i(t) = \eta_i m_{dir}(t) + [1 + 0.82 \log K_i(t)][0.09k_i(t) + 0.19k_i(t-1)] \qquad (A.2)$$

is more satisfactory. This means that to approximately predict the number of citations that a paper garners in the year $t+1$ after publication, in most cases it is enough to know its citation history during previous couple of years: t and $t-1$. Thus our results validate the widespread use of the 2-year impact factor.

References

1. Adams, J. (2005). Early citation counts correlate with accumulated impact. *Scientometrics, 63*(3), 567–581.
2. Albarrán, P., Crespo, J. A., Ortuño, I., & Ruiz-Castillo, J. (2011). The skewness of science in 219 sub-fields and a number of aggregates. *Scientometrics, 88*(2), 385–397.
3. Albert, R., & Barabasi, A. L. (2002). Statistical mechanics of complex networks. *Reviews of Modern Physics, 74*, 47–97.
4. Auerbach, F. (1913). Das Gesetz der Bevölkerungskonzentration. *Petermanns Geographische Mittelungen 59*, 74–76.
5. Bagrow, J. P., & Brockmann, D. (2013). Natural emergence of clusters and bursts in network evolution. *Physical Review X, 3*, 021016.
6. Bar-Ilan, J. (2008). Informetrics at the beginning of the 21st century–a review. *Journal of Informetrics, 2*(1), 1–52.
7. Barabasi, A. L. (2015). *Network science*. Cambridge: Cambridge University Press.
8. Barabasi, A. L., & Albert R. (1999). Emergence of scaling in random networks. *Science, 286*(5439), 509–512.
9. Barabasi, A. L., Song, C., & Wang D. (2012). Publishing: Handful of papers dominates citation. *Nature, 491*(7422), 40.
10. Baumgartner, S. E., & Leydesdorff L. (2013). Group-based trajectory modeling (GBTM) of citations in scholarly literature: Dynamic qualities of "transient" and "sticky knowledge claims". *Journal of the Association for Information Science and Technology, 65*(4), 797–811.
11. Bedogne, C., & Rodgers, G. J. (2006). Complex growing networks with intrinsic vertex fitness. *Physical Review E, 74*(4), 046115.
12. Bell, M., Perera, S., Piraveenan, M., Bliemer, M., Latty, T., & Reid, C. (2017). Network growth models: A behavioural basis for attachment proportional to fitness. *Scientific Reports, 7*, 42431.
13. Bertin, M., Atanassova, I., Gingras, Y., & Larivière, V. (2015). The invariant distribution of references in scientific articles. *Journal of the Association for Information Science and Technology, 67*(1), 164–177.
14. Bianconi, G., & Barabasi, A.-L. (2001). Bose-Einstein condensation in complex networks. *Physical Review Letters, 86*, 5632–5635.
15. Boccaletti, S., Latora, V., Moreno, Y., Chavez, M., & Hwang, D.-U. (2006). Complex networks: Structure and dynamics. *Physics Reports, 424*(4–5), 175–308.
16. Bollen, J., Van de Sompel, H., Hagberg, A., & Chute, R. (2009). A principal component analysis of 39 scientific impact measures. *PLoS One, 4*(6), e6022.

© The Author(s), under exclusive license to Springer Nature Switzerland AG 2019   109
M. Golosovsky, *Citation Analysis and Dynamics of Citation Networks*,
SpringerBriefs in Complexity, https://doi.org/10.1007/978-3-030-28169-4

17. Bornmann, L., & Daniel, H.-D. (2009). Universality of citation distributions—A validation of Radicchi et al.'s relative indicator $c_f = c/c_0$ at the micro level using data from chemistry. *Journal of the American Society for Information Science and Technology, 60*(8), 1664–1670.

18. Bornmann, L., & Haunschild, R. (2016). Citation score normalized by cited references (CSNCR): The introduction of a new citation impact indicator. *Journal of Informetrics, 10*(3), 875–887.

19. Bradford, S. C. (1934). Sources of information on specific subjects, *Engineering: An Illustrated Weekly Journal (London), 137*, 85–86. Reprinted as: Bradford, S. C. (1985). Sources of information on specific subjects. *Journal of Information Science, 10*(4), 176–180.

20. Bramoullé, Y., Currarini, S., Jackson, M. O., Pin, P., & Rogers, B. W. (2012). Homophily and long-run integration in social networks. *Journal of Economic Theory, 147*(5), 1754–1786.

21. Broder, A., Kumar, R., Maghoul, F., Raghavan, P., Rajagopalan, S., Stata, R., et. al. (2000). Graph structure in the web. *Computer Networks, 33*(1–6), 309–320.

22. Broido, A. D., & Clauset, A. (2019). Scale-free networks are rare. *Nature Communications, 10*(1), 1017.

23. Brot, H., & Louzoun, Y. Private communication.

24. Brzezinski, M. (2015). Power laws in citation distributions: evidence from Scopus. *Scientometrics, 103*(1), 213–228.

25. Burrell, Q. L. (2003). Predicting future citation behavior. *Journal of the American Society for Information Science and Technology, 54*(5), 372–378.

26. Burrell, Q. L. (2005). The use of the generalized Waring process in modelling informetric data. *Scientometrics, 64*(3), 247–270.

27. Burrell, Q. L. (2013). A stochastic approach to the relation between the impact factor and the uncitedness factor. *Journal of Informetrics, 7*(3), 676–682.

28. Caldarelli, G. (2007). *Scale-free networks: Complex webs in nature and technology*. Oxford: Oxford University Press.

29. Caldarelli, G., Capocci, A., De Los Rios, P., & Muñoz, M. A. (2002). Scale-free networks from varying vertex intrinsic fitness. *Physical Review Letters, 89*(25), 258702.

30. Candia, C., Jara-Figueroa, C., Rodriguez-Sickert, C., Barabási, A. L., & Hidalgo, C. A. (2019). The universal decay of collective memory and attention. *Nature Human Behaviour, 3*(1), 82–91.

31. Cao, X., Chen, Y., & Ray Liu, K. J. (2016). A data analytic approach to quantifying scientific impact. *Journal of Informetrics, 10*(2), 471–484.

32. Capocci, A., Servedio, V. D., Colaiori, F., Buriol, L. S., Donato, D., Leonardi, S., et al. (2006). Preferential attachment in the growth of social networks: The internet encyclopedia Wikipedia. *Physical Review E, 74*(3), 036116.

33. Carletti, T., Gargiulo, F., & Lambiotte, R. (2015). Preferential attachment with partial information. *The European Physical Journal B, 88*(1), 18.

34. Castillo, C., Donato, D., & Gionis, A. (2007). Estimating number of citations using author reputation. In N. Ziviani & R. Baeza-Yates (Eds.), *String processing and information retrieval* (pp. 107–117). Berlin: Springer.

35. Centola, D. (2010). The spread of behavior in an online social network experiment. *Science, 329*(5996), 1194–1197.

36. Centola, D., Eguíluz, V. M., & Macy, M. W. (2007). Cascade dynamics of complex propagation. *Physica A: Statistical Mechanics and Its Applications, 374*(1), 449–456.

37. Chakrabarti, B. K.,& Sen, P. (2014). *Sociophysics: An introduction*. Oxford: Oxford University Press.

38. Chatterjee, A., Ghosh, A., & Chakrabarti, B. K. (2014). Universality of citation distributions for academic institutions and journals. *PLoS One, 11*, e0146762.

39. Ciotti, V., Bonaventura, M., Nicosia, V., Panzarasa, P., & Latora, V. (2016). Homophily and missing links in citation networks. *EPJ Data Science, 5*(1), 7.

40. Clauset, A., Larremore, D. B., & Sinatra, R. (2017). Data-driven predictions in the science of science. *Science, 355*(6324), 477–480.

41. Clauset, A., Shalizi, C., & Newman, M. (2009). Power-law distributions in empirical data. *SIAM Review, 51*(4), 661–703.
42. Clough, J. R., Gollings, J., Loach, T. V., & Evans, T. S. (2014). Transitive reduction of citation networks. *Journal of Complex Networks, 3*(2), 189–203.
43. Condon, E. U. (1928). Statistics of vocabulary. *Science, 67*(1733), 300.
44. Csárdi, G., Strandburg, K. J., Zalányi, L., Tobochnik, J., & Érdi, P. (2007). Modeling innovation by a kinetic description of the patent citation system. *Physica A: Statistical Mechanics and Its Applications, 374*(2), 783–793.
45. de Solla Price, D. J. (1965). Networks of scientific papers. *Science, 149*(3683), 510–515.
46. Didegah, F., & Thelwall, M. (2013). Which factors help authors produce the highest impact research? Collaboration, journal and document properties. *Journal of Informetrics, 7*(4), 861–873.
47. Dorogovtsev, S. N., & Mendes, J. F. F. (2000). Evolution of networks with aging of sites. *Physical Review E, 62*(2), 1842–1845.
48. Dorogovtsev, S. N., & Mendes, J. F. F. (2001). Scaling properties of scale-free evolving networks: Continuous approach. *Physical Review E, 63*(5), 056125.
49. Eom, Y.-H., & Fortunato, S. (2011). Characterizing and modeling citation dynamics. *PLoS One, 6*(9), e24926.
50. Ergün, G., & Rodgers, G. J. (2002). Growing random networks with fitness. *Physica A: Statistical Mechanics and Its Applications, 303*(1), 261–272.
51. Evans, T. S., Hopkins, N., & Kaube, B. S. (2012). Universality of performance indicators based on citation and reference counts. *Scientometrics, 93*(2), 473–495.
52. Fortunato, S. (2010). Community detection in graphs. *Physics Reports, 486*(3), 75–174.
53. Fortunato, S., Bergstrom, C. T., Börner, K., Evans, J. A., Helbing, D., Milojević, S., et al. (2018). Science of science. *Science, 359*(6379), eaao0185.
54. Fortunato, S., Flammini, A., & Menczer, F. (2006). Scale-free network growth by ranking. *Physical Review Letters, 96*(21), 218701.
55. Fox, C. W., Timothy Paine, C. E., & Sauterey, B. (2019). Citations increase with manuscript length, author number, and references cited in ecology journals. *Ecology and Evolution, 6*(21), 7717–7726.
56. Gao, J., Barzel, B., & Barabási, A. L. (2016). Universal resilience patterns in complex networks. *Nature, 530*, 307.
57. Garavaglia, A., Hofstad, R., & Woeginger, G. (2017). The dynamics of power laws: Fitness and aging in preferential attachment trees. *Journal of Statistical Physics, 168*(6), 1137–1179.
58. Garfield, E. (1998). Random thoughts on citationology its theory, & practice. *Scientometrics, 43*(1), 69–76.
59. Geller, N. L., de Cani, J. S., & Davies, R. E. (1981). Lifetime-citation rates: A mathematical model to compare scientists' work. *Journal of the American Society for Information Science, 32*(1), 1–15.
60. Geng, X., & Wang, Y. (2009). Degree correlations in citation networks model with aging. *Europhysics Letters, 88*(3), 38002.
61. Ghadge, S., Killingback, T., Sundaram, B., & Tran, D. A. (2010). A statistical construction of power-law networks. *International Journal of Parallel, Emergent and Distributed Systems, 25*(3), 223–235.
62. Gittes, F. T., & Schick, M. (1984). Complete and incomplete wetting by adsorbed solids. *Physical Review B, 30*(1), 209.
63. Glanzel, W. (2004). Towards a model for diachronous and synchronous citation analyses. *Scientometrics, 60*(3), 511–522.
64. Glanzel, W., Schlemmer, B., & Thijs, B. (2003). Better late than never? On the chance to become highly cited only beyond the standard bibliometric time horizon. *Scientometrics, 58*(3), 571–586.
65. Gleeson, J. P., Cellai, D., Onnela, J.-P., Porter, M. A., & Reed-Tsochas, F. (2014). A simple generative model of collective online behavior. *Proceedings of the National Academy of Sciences, 111*(29), 10411–10415.

66. Goldberg, S. R., Anthony, H., & Evans, T. S. (2015). Modelling citation networks. *Scientometrics, 105*(3), 1577–1604.
67. Golosovsky, M. (2017). Power-law citation distributions are not scale-free. *Physical Review E, 96*(3), 032306.
68. Golosovsky, M. (2018). Mechanisms of complex network growth: Synthesis of the preferential attachment and fitness models. *Physical Review E, 97*(6), 062310.
69. Golosovsky, M., & Solomon, S. (2012). Runaway events dominate the heavy tail of citation distributions. *The European Physical Journal, 205*(1), 303–311.
70. Golosovsky, M., & Solomon, S. (2012). Stochastic dynamical model of a growing citation network based on a self-exciting point process. *Physical Review Letters, 109*(9), 098701.
71. Golosovsky, M., & Solomon, S. (2013). The transition towards immortality: Non-linear autocatalytic growth of citations to scientific papers. *Journal of Statistical Physics, 151*(1–2), 340–354.
72. Golosovsky, M., & Solomon, S. (2017). Growing complex network of citations of scientific papers: Modeling and measurements. *Physical Review E, 95*(1), 012324.
73. Hajra, K. B., & Sen, P. (2006). Modelling aging characteristics in citation networks. *Physica A: Statistical Mechanics and Its Applications, 368*(2), 575–582.
74. Hazouglou, M. J., Kulkarni, V., Skiena, S. S., & Dill, K. A. (2017). Citation histories of papers: sometimes the rich get richer, sometimes they don't. *Preprint arXiv:1703.04746*.
75. Higham, K. W., Governale, M., Jaffe, A. B., & Zülicke, U. (2019). Ex-ante measure of patent quality reveals intrinsic fitness for citation-network growth. *Physical Review E, 99*(6), 060301.
76. Higham, K. W., Governale, M., Jaffe, A. B., & Zülicke, U. (2017). Fame and obsolescence: Disentangling growth and aging dynamics of patent citations. *Physical Review E, 95*(4), 042309.
77. Higham, K. W., Governale, M., Jaffe, A. B., & Zülicke, U. (2017). Unraveling the dynamics of growth, aging and inflation for citations to scientific articles from specific research fields. *Journal of Informetrics, 11*(4), 1190–1200.
78. Himpsel, F. J., Marcus, P. M., Tromp, R., Batra, I. P., Cook, M. R., Jona, F., et al. (1984). Structure analysis of Si (111) 2x1 with low-energy electron diffraction. *Physical Review B, 30*(4), 2257–2259.
79. Hirsch, J. E. (2005). An index to quantify an individual's scientific research output. *Proceedings of the National Academy of Sciences, 102*(46), 16569–16572.
80. Hsu, J. W., & Huang, D. W. (2012). A scaling between impact factor and uncitedness. *Physica A: Statistical Mechanics and Its Applications, 391*(5), 2129–2134.
81. Huberman, B. A., & Adamic, L. A. (1999). Evolutionary dynamics of the World Wide Web. *Preprint arXiv:cond-mat/9901071*.
82. Jackson, M. O., & Rogers, B. W. (2007). Meeting strangers and friends of friends: How random are social networks? *American Economic Review, 97*(3), 890–915.
83. Jeong, H., Néda, Z., & Barabási, A.-L. (2003). Measuring preferential attachment in evolving networks. *Europhysics Letters, 61*(4), 567–572.
84. Ke, Q., Ferrara, E., Radicchi, F., & Flammini, A. (2015). Defining and identifying sleeping beauties in science. *Proceedings of the National Academy of Sciences, 112*(24), 7426–7431.
85. Ke, W. (2013). A fitness model for scholarly impact analysis. *Scientometrics, 94*(3), 981–998.
86. Klimek, P., Jovanovic, A. S., Egloff, R., & Schneider, R. (2016). Successful fish go with the flow: Citation impact prediction based on centrality measures for term–document networks. *Scientometrics, 107*(3), 1265–1282.
87. Kong, J. S., Sarshar, N., & Roychowdhury, V. P. (2008). Experience versus talent shapes the structure of the Web. *Proceedings of the National Academy of Sciences, 105*(37), 13724–13729.
88. Krapivsky, P. L., & Redner, S. (2001). Organization of growing random networks. *Physical Review E, 63*(6), 066123.
89. Krapivsky, P. L., & Redner, S. (2005). Network growth by copying. *Physical Review E, 71*, 036118.

90. Kunegis, J., Blattner, M., & Moser, C. (2013). Preferential attachment in online networks. In *Proceedings of the 5th Annual ACM Web Science Conference*. New York, NY: Association for Computing Machinery.

91. Lambiotte, R., & Ausloos, M. (2007). Growing network with j-redirection. *Europhysics Letters, 77*(5), 58002.

92. Lambiotte, R., Krapivsky, P. L., Bhat, U., & Redner, S. (2016). Structural transitions in densifying networks. *Physical Review Letters, 117*(21), 218301.

93. Larivière, V., & Gingras, Y. (2010). The impact factor's Matthew effect: A natural experiment in bibliometrics. *Journal of the Association for Information Science & Technology, 61*(2), 424–427.

94. Lehmann, S., Jackson, A. D., & Lautrup, B. (2005). Life, death and preferential attachment. *Europhysics Letters, 69*(2), 298–303.

95. Leskovec, J., Backstrom, L., Kumar, R., & Tomkins, A. (2008). Microscopic evolution of social networks. In *Proceeding of the 14th ACM SIGKDD International Conference on Knowledge Discovery and Data Mining - KDD 08*. New York, NY: Association for Computing Machinery.

96. Leskovec, J., Kleinberg, J., & Faloutsos, C. (2005). Graphs over time. In *Proceeding of the Eleventh ACM SIGKDD International Conference on Knowledge Discovery in Data Mining - KDD 05*. New York, NY: Association for Computing Machinery.

97. Leskovec, J., Kleinberg, J., & Faloutsos, C. (2007). Graph evolution: Densification and shrinking diameters. *ACM Transactions on Knowledge Discovery from Data (TKDD), 1*(1), 2.

98. Letchford, A., Moat, H. S., & Preis, T. (2015). The advantage of short paper titles. *Royal Society Open Science, 2*(8), 150266.

99. Leydesdorff, L., Wagner, C. S., & Bornmann, L. (2018). Discontinuities in citation relations among journals: Self-organized criticality as a model of scientific revolutions and change. *Scientometrics, 116*(1), 623–644.

100. Li, L., & Tong, H. (2015). The child is father of the man: Foresee the success at the early stage. In *Proceedings of the 21th ACM SIGKDD International Conference on Knowledge Discovery and Data Mining*. New York, NY: ACM.

101. Liben-Nowell, D., & Kleinberg, J. (2003). The link prediction problem for social networks. In *Proceedings of the Twelfth International Conference on Information and Knowledge Management - CIKM*. New York, NY: Association for Computing Machinery.

102. Limpert, E., Stahel, W. A., & Abbt, M. (2001). Log-normal distributions across the sciences: Keys and clues. *BioScience, 51*(5), 341–352.

103. Lotka, A. J. (1926). The frequency distribution of scientific productivity. *Journal of the Washington Academy of Sciences, 16*(11), 317–323.

104. Luck, J. M., & Mehta, A. (2017). How the fittest compete for leadership: A tale of tails. *Physical Review E, 95*(6), 062306.

105. Mariani, M. S., Medo, M., & Lafond, F. (2019). Early identification of important patents: Design and validation of citation network metrics. *Technological Forecasting and Social Change, 146*, 644–654.

106. Martin, T., Ball, B., Karrer, B., & Newman, M. E. J. (2013). Coauthorship and citation patterns in the physical review. *Physical Review E, 88*(1), 012814.

107. Medo, M., Cimini, G., & Gualdi, S. (2011). Temporal effects in the growth of networks. *Physical Review Letters, 107*, 238701.

108. Menczer, F. (2004). Evolution of document networks. *Proceedings of the National Academy of Sciences, 101*(Supplement 1), 5261–5265.

109. Miller, B. A., & Bliss, N. T. (2012). A stochastic system for large network growth. *IEEE Signal Processing Letters, 19*(6), 356–359.

110. Min, C., Ding, Y., Li, J., Bu, Y., Pei, L., & Sun, J. (2018). Innovation or imitation: The diffusion of citations. *Journal of the Association for Information Science and Technology, 69*(10), 1271–1282.

111. Mingers, J., & Burrell, Q. L. (2006). Modeling citation behavior in Management Science journals. *Information Processing & Management, 42*(6), 1451–1464.
112. Mingers, J., & Leydesdorff, L. (2015). A review of theory, & practice in scientometrics. *European Journal of Operational Research, 246*(1), 1–19.
113. Mislove A., Koppula H. S., Gummadi K. P., Druschel P., & Bhattacharjee B. (2013). An empirical validation of growth models for complex networks. In A. Mukherjee, M. Choudhury, F. Peruani, N. Ganguly, & B. Mitra (Eds.), *Dynamics on and of complex networks. Modeling and simulation in science, engineering and technology* (Vol. 2). New York, NY: Birkhäuser.
114. Mitzenmacher, M. (2004). A brief history of generative models for power law and lognormal distributions. *Internet Mathematics, 1*(2), 226–251.
115. Mitzenmacher, M. (2005). Editorial: The future of power law research. *Internet Mathematics, 2*(4), 525–534.
116. Moed, H. F. (2005). *Citation analysis in research evaluation*. Berlin: Springer.
117. Mokryn, O., & Reznik, A. (2015). On skewed distributions and straight lines. In *Proceedings of the 24th International Conference on World Wide Web*. New York, NY: Association for Computing Machinery.
118. Mokryn, O., Wagner, A., Blattner, M., Ruppin, E., & Shavitt, Y. (2016). The role of temporal trends in growing networks. *PLoS One, 11*(8), e0156505.
119. Mukherjee, S., Romero, D. M., Jones, B., & Uzzi, B. (2017). The nearly universal link between the age of past knowledge and tomorrows breakthroughs in science and technology: The hotspot. *Science Advances, 3*(4), e1601315.
120. Nakamoto, H. (1988). Synchronous and diachronous citation distributions. *Informetrics, 87/88*, 157–163.
121. Newman, M. (2010). *Networks*. Oxford: Oxford University Press.
122. Newman, M. E. (2001). Clustering and preferential attachment in growing networks. *Physical Review E, 64*(2), 025102.
123. Newman, M. E. (2005). Power Laws, Pareto distributions and Zipf's law. *Contemporary Physics, 46*(5), 323–351.
124. Newman, M. E. J. (2009). The first-mover advantage in scientific publication. *Europhysics Letters, 86*(6), 68001.
125. Newman, M. E. J. (2014). Prediction of highly cited papers. *Europhysics Letters, 105*(2), 28002.
126. Nguyen, K., & Tran, D. A. (2012). Fitness-based generative models for power-law networks. In *Handbook of optimization in complex networks* (pp. 39–53). Berlin: Springer.
127. Ostroumova-Prokhorenkova, L., & Samosvat, E. (2016). Recency-based preferential attachment models. *Journal of Complex Networks, 4*(4), 475–499.
128. Papadopoulos, F., Kitsak, M., Serrano, M. Á., Boguñá, M., & Krioukov, D. (2012). Popularity versus similarity in growing networks. *Nature, 489*(7417), 537–540.
129. Pareto, V. (1909). *Manuel d'economie politique*. Paris: Giard & Briére.
130. Pennock, D. M., Flake, G. W., Lawrence, S., Glover, E. J., & Giles, C. L. (2002). Winners don't take all: Characterizing the competition for links on the web. *Proceedings of the National Academy of Sciences, 99*(8), 5207–5211.
131. Perc, M. (2014).The Matthew effect in empirical data. *Journal of the Royal Society Interface, 11*, 20140378.
132. Peters, H. P. F., & van Raan, A. F. J. (1994). On determinants of citation scores: A case study in chemical engineering. *Journal of the American Society for Information Science, 45*(1), 39–49.
133. Peterson, G. J., Presse, S., & Dill, K. A. (2010). Nonuniversal power law scaling in the probability distribution of scientific citations. *Proceedings of the National Academy of Sciences, 107*(37), 16023–16027.
134. Pham, T., Sheridan, P., & Shimodaira, H. (2015). PAFit: A statistical method for measuring preferential attachment in temporal complex networks. *PLoS One, 10*(9), e0137796.

135. Pham, T., Sheridan, P., & Shimodaira, H. (2016). Joint estimation of preferential attachment and node fitness in growing complex networks. *Scientific Reports, 6*, 32558.
136. Pinto, C. M. A., Lopes, A. M., & Machado, J. A. T. (2012). A review of power laws in real life phenomena. *Communications in Nonlinear Science and Numerical Simulation, 17*(9), 3558–3578.
137. Ponomarev, I. V., Lawton, B. K., Williams, D. E., & Schnell, J. D. (2014). Breakthrough paper indicator 2.0: Can geographical diversity and interdisciplinarity improve the accuracy of outstanding papers prediction? *Scientometrics, 100*(3), 755–765.
138. Price, D. D. S. (1976). A general theory of bibliometric and other cumulative advantage processes. *Journal of the American Society for Information Science, 27*(5), 292–306.
139. Radicchi, F., & Castellano, C. (2011). Rescaling citations of publications in physics. *Physical Review E, 83*(4), 046116.
140. Radicchi, F., & Castellano, C. (2012). A reverse engineering approach to the suppression of citation biases reveals universal properties of citation distributions. *PLoS One, 7*(3), e33833.
141. Radicchi, F., Fortunato, S., & Castellano, C. (2008). Universality of citation distributions: Toward an objective measure of scientific impact. *Proceedings of the National Academy of Sciences, 105*(45), 17268–17272.
142. Redner, S. (1998). How popular is your paper? An empirical study of the citation distribution. *The European Physical Journal B, 4*(2), 131–134.
143. Redner, S. (2004). Citation statistics from more than a century of Physical Review. *Preprint arXiv:physics/0407137.*
144. Redner, S. (2005). Citation statistics from 110 years of Physical Review. *Physics Today, 58*(6), 49–54.
145. Rosvall, M., Esquivel, A. V., Lancichinetti, A., West, J. D., & Lambiotte, R. (2014). Memory in network flows and its effects on spreading dynamics and community detection. *Nature Communications, 5*, 4630.
146. Roth, C., Wu, J., & Lozano, S. (2012). Assessing impact and quality from local dynamics of citation networks. *Journal of Informetrics, 6*(1), 111–120.
147. Scharnhorst, A., Börner, K., & van den Besselaar, P. A. A. (2012). *Models of science dynamics; Encounters between complexity theory and information sciences.* Understanding complex systems. Berlin: Springer.
148. Seglen, P. O. (1992). The skewness of science. *Journal of the American Society for Information Science, 43*(9), 628–638.
149. Sendiña-Nadal, I., Danziger, M. M., Wang, Z., Havlin, S., & Boccaletti, S. (2016). Assortativity and leadership emerge from anti-preferential attachment in heterogeneous networks. *Scientific Reports, 6*(1), 21297.
150. Servedio, V. D. P., Caldarelli, G., & Buttà, P. (2004). Vertex intrinsic fitness: How to produce arbitrary scale-free networks. *Physical Review E, 70*(5), 056126.
151. Shao, Z. G., Zou, X. W., Tan, Z. J., & Jin, Z. Z. (2006). Growing networks with mixed attachment mechanisms. *Journal of Physics A: Mathematical and General, 39*(9), 2035–2042.
152. Shih, W. Y., & Stroud, D. (1984). Superconducting arrays in a magnetic field: Effects of lattice structure and a possible double transition. *Physical Review B, 30*(11), 6774.
153. Shockley, W. (1957). On the statistics of individual variations of productivity in research laboratories. *Proceedings of the IRE, 45*(3), 279–290.
154. Simkin, M. V., & Roychowdhury, V. P. (2007). A mathematical theory of citing. *Journal of the American Society for Information Science and Technology, 58*(11), 1661–1673.
155. Sinatra, R., Deville, P., Szell, M., Wang, D., & Barabási, A. L. (2015). A century of physics. *Nature Physics, 11*:791.
156. Small, H. (1973). Co-citation in the scientific literature: A new measure of the relationship between two documents. *Journal of the American Society for information Science, 24*(4), 265–269.
157. Solomon, S., Yaari, G., Dover, Y., & Moulet, S. (2009). Do all economies grow equally fast? *Risk and Decision Analysis, 1*(3), 171–185.

158. Sornette, D. (2012). Probability distributions in complex systems. In *Computational complexity* (pp. 2286–2300). New York, NY: Springer.
159. Stegehuis, C., Litvak, N., & Waltman, L. (2015). Predicting the long-term citation impact of recent publications. *Journal of Informetrics, 9*(3), 642–657.
160. Stringer, M. J., Sales-Pardo, M., & Amaral, L. A. N. (2008). Effectiveness of journal ranking schemes as a tool for locating information. *PLoS One, 3*(2), e1683.
161. Stumpf, M. P. H., & Porter, M. A. (2012). Critical truths about power laws. *Science, 335*(6069), 665–666.
162. Šubelj, L., & Bajec, M. (2013). Model of complex networks based on citation dynamics. In *Proceedings of the WWW Workshop on Large Scale Network Analysis, 2013:(LSNA'13)* (pp.527–530).
163. Sugimoto, C. R., & Cronin, B. (2014). *Beyond bibliometrics: Harnessing multidimensional indicators of scholarly impact*. Cambridge: The MIT Press.
164. Tahamtan, I., Afshar, A. S., & Ahamdzadeh, K. (2016). Factors affecting number of citations: a comprehensive review of the literature. *Scientometrics, 107*(3), 1195–1225.
165. Thelwall, M. (2016). The discretised lognormal and hooked power law distributions for complete citation data: Best options for modelling and regression. *Journal of Informetrics, 10*(2), 336–346.
166. Šubelj, L., & Fiala, D. (2017). Publication boost in Web of Science journals and its effect on citation distributions. *Journal of the Association for Information Science and Technology, 68*(4), 1018–1023.
167. Uzzi, B., Mukherjee, S., Stringer, M., & Jones, B. (2013). Atypical combinations and scientific impact. *Science, 342*(6157), 468–472.
168. Van Noorden, R. (2017). The science that's never been cited. *Nature, 552,* 162–164.
169. van Raan, A. F. J. (2004). Sleeping beauties in science. *Scientometrics, 59*(3), 467–472.
170. Vazquez, A. (2001). Disordered networks generated by recursive searches. *Europhysics Letters, 54*(4), 430–435.
171. Vazquez, A. (2001). Statistics of citation networks. *Preprint arXiv:cond-mat/0105031.*
172. Vazquez, A. (2003). Growing network with local rules: Preferential attachment, clustering hierarchy, and degree correlations. *Physical Review E, 67,* 056104.
173. Waltman, L. (2016). A review of the literature on citation impact indicators. *Journal of Informetrics, 10*(2), 365–391.
174. Waltman, L., van Eck, N. J., & van Raan, A. F. (2012). Universality of citation distributions revisited. *Journal of the American Society for Information Science and Technology, 63*(1), 72–77.
175. Wang, D., Song, C., & Barabási, A. L. (2013). Quantifying long-term scientific impact. *Science, 342*(6154), 127–132.
176. Wang, M., Yu, G., & Yu, D. (2008). Measuring the preferential attachment mechanism in citation networks. *Physica A: Statistical Mechanics and Its Applications, 387*(18), 4692–4698.
177. Willinger, W., Alderson, D., & Doyle, J. C. (2009). Mathematics and the Internet: A source of enormous confusion and great potential. *Notices of the AMS, 56*(5), 586–599.
178. Wu, Y., Fu, T. Z. J., & Chiu, D. M. (2014). Generalized preferential attachment considering aging. *Journal of Informetrics, 8*(3), 650–658.
179. Wu, Z.-X., & Holme, P. (2009). Modeling scientific-citation patterns and other triangle-rich acyclic networks. *Physical Review E, 80,* 037101.
180. Xie, Z., Ouyang, Z., Liu, Q., & Li, J. (2016). A geometric graph model for citation networks of exponentially growing scientific papers. *Physica A: Statistical Mechanics and Its Applications, 456,* 167–175.
181. Yan, R., Huang, C., Tang, J., Zhang, Y., & Li, X. (2012). To better stand on the shoulder of giants. In *Proceedings of the 12th ACM/IEEE-CS Joint Conference on Digital Libraries, JCDL '12* (pp. 51–60). New York, NY: ACM.

182. Yan, R., Tang, J., Liu, X., Shan, D., & Li, X. (2011). Citation count prediction: Learning to estimate future citations for literature. In *Proceedings of the 20th ACM International Conference on Information and Knowledge Management, CIKM '11* (pp. 1247–1252). New York, NY: ACM.

183. Yin, Y., & Wang, D. (2017). The time dimension of science: Connecting the past to the future. *Journal of Informetrics, 11*(2), 608–621.

184. Zener, C. (1968). An analysis of scientific productivity. *Proceedings of the National Academy of Sciences of the United States of America, 59*(4), 1078–1081.

185. Zeng, A., Shen, Z., Zhou, J., Wu, J., Fan, Y., Wang, Y., & Stanley, H. E. (2017). The science of science: From the perspective of complex systems. *Physics Reports, 714–715*, 1–73.

186. Zhou, S., & Mondragon. R. J. (2004). Accurately modeling the internet topology. *Physical Review E, 70*, 066108.

187. Zhou, J., Zeng, A., Fan, Y., & Di, Z. (2016) Ranking scientific publications with similarity-preferential mechanism. *Scientometrics, 106*(2), 805–816.

188. Zipf, G. K. (1949) *Human behaviour and the principle of least effort*. Reading, MA: Addison-Wesley.

Index

© The Author(s), under exclusive license to Springer Nature Switzerland AG 2019　　119
M. Golosovsky, *Citation Analysis and Dynamics of Citation Networks*,
SpringerBriefs in Complexity, https://doi.org/10.1007/978-3-030-28169-4

Printed in the United States
By Bookmasters